Managing Salinization

Springer

Berlin
Heidelberg
New York
Barcelona
Budapest
Hong Kong
London
Milan
Paris
Santa Clara
Singapore
Tokyo

Waltina Scheumann

Managing Salinization

Institutional Analysis
of Public Irrigation Systems

With 18 Figures and 36 Tables

 Springer

Dr.-Ing. Waltina Scheumann
Diplom-Politologin
Lefèvrestraße 22
D-12161 Berlin

This work was accepted 1996 by Department 7 – Environment and Society of the Technical University of Berlin (Fachbereich 7 – Umwelt und Gesellschaft der Technischen Universität Berlin) under the title *Institutional Analysis of Salinization in Public Irrigation Systems* as dissertation for a doctoral thesis.

Cataloging-in-Publication Data applied for.

Die Deutsche Bibliothek – CIP-Einheitsaufnahme

Scheumann, Waltina:
Managing salinization: institutional analysis of public irrigation systems; with 36 tables / Waltina Scheumann. – Berlin; Heidelberg; New York; Barcelona; Budapest; Hong Kong; London; Milan; Paris, Santa Clara; Singapore; Tokyo: Springer, 1997
ISBN 3-540-63328-6

ISBN 3-540-63328-6 Springer-Verlag Berlin Heidelberg New York

Cover design: de'blik, Berlin
Typesetting: Camera-ready by the author
SPIN 10566155   41/3136 – 5 4 3 2 1 0 – Printed on acid-free paper

Contents

Chapter Four: Contributions to High Groundwater Levels and Salinization Caused by the Operation and Maintenance of the Lower Seyhan Irrigation Project 131

Chapter Five: Effects of High Groundwater Levels, Waterlogging and Salinity on Farm Economy 181

List of illustrations

Maps

Tables

X   Contents

Boxes

Abbreviations

DSI	the Turkish acronym for the General Directorate for State Hydraulic Works (Devlet Su Isleri Genel Müdürlügü)
EC	Electrical Conductivity of water is the reciprocal of electrical resistance measured at a temperature of 25 $^{\circ}$C (micromhos/cm, or mmhos/cm). EC_e is used for the electrical conductivity of the soil saturation extract, EC_i for that of irrigation water, EC_d for drainage water, and EC_{gw} for groundwater.
FAO	Food and Agriculture Organization of the United Nations
GAP	the Turkish acronym for the Southeastern Anatolian Project (Güneydogu Anadolu Projesi)
GDRS	the English acronym for the General Directorate for Rural Services (Köy Hizmetleri Genel Müdürlügü)
IBRD	International Bank for Reconstruction and Development
ICID	International Commission on Irrigation and Drainage
IDA	International Development Association
IIMI	International Irrigation Management Institute
IRC	Institutional Rational Choice approach
KTK	the Turkish acronym for Water User Groups which are based on the villages' administrative units (Köy Tüzel Kisiligi)
MoF	Ministry of Finance
NIA	National Irrigation Administration of the Philippines
NIE	New Institutional Economics
O&M	Operation&Maintenance
SAR	Sodium-Adsorption Rate is a ratio for soil extracts and irrigation waters used to express the relative activity of sodium ions in the exchange reactions with soil.
SPO	State Planning Organization
TAMS	the acronym for the consultant Tippetts-Abbett-McCarthy-Stratton
TD or YD	main drainage canals in the Yüregir and Tarsus Plain
TL	Turkish Lira
TOPRAKSU	the Turkish acronym for the General Directorate for Soil and Water (Toprak Su Genel Müdürlügü)
TP or YP	Tarsus or Yüregir Plain; also used to note subunits within the plain
TS or YS	main delivery canals in the Yüregir and Tarsus Plain
WUA	Water User Association
WUG	Water User Group

Zusammenfassung

Obwohl die technischen Lösungen seit Jahrzehnten bekannt sind, gehören hohe Grundwasserstände und die Versalzung landwirtschaftlicher Flächen nach wie vor zu den häufigsten und gravierendsten Umweltproblemen, die in großen staatlichen Bewässerungsprojekten in ariden und semiariden Gebieten auftreten. Die von dem Forschungsansatz der Neuen Institutionenökonomie geleitete vorliegende Studie geht davon aus, daß 'Institutionen' für den effizienten Gebrauch von Ressourcen, für die Internalisierung externer Effekte und für die Bereitstellung der Dienstleistung 'Versalzungskontrolle' entscheidend sind. Institutionen werden als sozial anerkannte Regeln und Konventionen verstanden, nach denen die Nutzer über die Ressourcen verfügen können.

In einem großen staatlichen Bewässerungsprojekt in der Südtürkei wird untersucht, inwiefern eine hochzentralisierte Entscheidungsstruktur Anreize schafft, die hohe Grundwasserstände und Versalzung verursachen, und ob die Mitwirkung von Wassernutzerorganisationen an Betrieb und Instandhaltung eines Bewässerungssystems die Bedingungen für eine effektive Versalzungskontrolle verbessern. Die staatliche Bewässerungsbehörde ist ein Monopolunternehmen, das keine zufriedenstellenden Dienstleistungen zur Versalzungskontrolle für die betroffenen Bauern bereitstellt. Da die Dienstleistungen von den politischen Entscheidungsträgern nachgefragt werden, unterliegen Entscheidungen über öffentliche Investitionen in Dränagesysteme und über die Instandhaltung der Be- und Entwässerungssysteme politischen Erwägungen und nicht lokalen und/oder technischen Erfordernissen. Die Budgetzuweisungen aus dem Staatshaushalt erfolgen unabhängig von der Qualität der Leistungen, und die direkten Nutznießer können weder direkt noch über ihre Kostenbeteiligung die Leistungen beeinflussen. Ihr Zahlungsbeitrag ist ohnehin minimal, da Freifahrerverhalten nicht effektiv ausgeschlossen werden kann. Die Höhe der eingezahlten Gebühren für Betrieb und Instandhaltung (O&M) hat wegen der institutionellen Struktur keinen Einfluß auf Investitionsentscheidungen und auf Entscheidungen über Art, Umfang und Qualität der Dienstleistungen. Die Untersuchung hat weiterhin gezeigt, daß hohe Transformations-, Informations- und Transaktionskosten eine Steuerung externer Effekte durch die staatliche Bewässerungsbehörde unmöglich oder ineffektiv machen. Die Beteiligung der Wassernutzerorganisationen hat positive Auswirkungen auf die Instandhaltung der Wasserzuleiter und auf die Durchsetzung der Wasserverteilungsregeln, die mit dem Grad ihrer Autonomie zunehmen. Allerdings ist ihre Beteiligung an der Implementierung von Maßnahmen zur Versalzungskontrolle minimal. In dem untersuchten Beispiel sind die negativen Auswirkungen der Versalzung auf die landwirtschaftlichen Betriebe marginal; die relativ kleine Anzahl von Betrieben, die unter sichtbaren Ertragseinbußen leiden, hat keine direkten Einflußmöglichkeiten auf Investitions- und Instandhaltungsentscheidungen.

'Managing Salinization' ist eine Herausforderung für die von der Neuen Institutionenökonomie geleitete Forschung: Die Untersuchung von Bewässerungssystemen, in denen eine signifikante Anzahl von Bauern von Versalzung betroffen ist zusammen mit den Konsequenzen für die Nutzerpartizipation steht aus, und systematische Untersuchungen über Anreizsysteme für Investitionen in Projekt- und Farmdräne und in ihre Instandhaltung fehlen.

Summary

For many decades, the general physical causes and the principal techniques for preventing and dealing with salinization have been well known and are well established. Despite these facts, however, high groundwater levels with or without the associated salinization of soils have continued to be widespread phenomena in irrigated agriculture counteracting the targets of the very costly public infrastructure investments, i.e., the increase in yields and incomes. Approaches are required which overcome the limits of technically oriented viewpoints. The New Institutional Economics (NIE) stress the importance of 'institutions' which may be characterized as socially accepted rules and conventions; they establish incentives, or not, for the multiple participants of a given institutional setting for an efficient allocation of resources, for the internalization of negative external effects and for the provision of adequate and effective salinization control services.

A large-scale public irrigation project in South Turkey with a highly centralized decision-making setting was selected for case study. In this state-regulated irrigation system, the irrigation agency is a monopoly which faces neither incentives for sustaining the infrastructure facilities or efficient water use nor for internalizing negative external effects. The agency receives instead its budget from central budget allocations, and the decisions made on the budget for operation and maintenance (O&M) are related not to the performance of the project, but to political consideration. In such a highly centralized decision-making setting, demand is expressed by political decision-makers not by the negatively affected farmers who have no direct means of control. New investments are favoured over the completion of older schemes, and the facilities are inadequately maintained; investments in drainage systems, and their maintenance, are neglected or poor. Public officials receive low salaries, and diligent services and good performance are not rewarded. Governmental control on the bureaucracy's service provision is weak, and the consumers/beneficiaries have an institutionalized influence on governmental decisions only through the mechanisms of the political system. Public officials are not accountable to them, and they are in no way dependent on those they serve, although irrigation is a service with consumers. The beneficiaries' financial contributions which are, however, negligible, have no influence on the services provided, because there is no linkage between the collected water charges and O&M budget allocations. In the analyzed case, the public provision of effective salinization control services is impeded by high transformation, information and transaction costs, and administrative procedures to minimize them themselves contribute to high groundwater levels. The case study has shown that with water-user organizations participating in O&M, maintenance works in the delivery systems and the enforcement of the water allocation rules have improved; their positive impacts increase with higher degrees of autonomy. However, their contribution to the control of salinization is minimal because the highly centralized decision-making setting continues to govern issues concerning the drainage systems. It was a peculiarity of the case study that the majority of the farmers were not negatively, or not too severely, affected by salinization. The small group of severely affected farmers had, however, no means of control over investment and maintenance decisions.

It would be useful to empirically evaluate cases where the majority of farmers are heavily affected, together with consequences for farmer participation and for the cooperation of water-user organizations with state agencies. Future research on salinization control would have to concentrate on two central issues; one refers to investments in drainage networks and their maintenance, and another deals with investments in on-farm drainage systems and their maintenance. Managing salinization is, however, a challenge to institutional analysts to elaborate new strategies.

Introduction

During the last four decades, irrigated agriculture has provided the major share of the increase in world crop production with 36% of the total crop production coming from less than 15% of the arable land that was irrigated. The annual expansion of irrigated land which made this positive development possible, increased by 1% in the early 1960s, and reached its maximum of 2.3% per year between 1972 and 1975; the rate then began to decline in the mid-1970s.[1] According to World Bank/ICID statistics, the irrigated area expanded from 1953 to 1970 by about 5 million ha per year; from 1970 to the early 1980s the rate still averaged about 4 million ha/year. Then, however, the rate of expansion declined and, since the early 1980s, has reached not more than 2 million ha per year.[2]

The reasons for this decline were the high capital costs of irrigation development which increased in the 1980s,[3] the low prices for food staples,[4] and diminishing land and water resources. The land which could easily be brought under irrigation was already developed, and the more expensive and economically less favourable areas were left. The same holds true for the development of water resources, which are, in addition, under high pressure from other sectors, because irrigated agriculture consumes by far most of the water used in almost all developing countries. In most countries of the Middle East and North Africa, for example, agriculture accounts for more than 85% of the water consumption.[5] Agriculture is, however, not only the world's largest water consumer in terms of volume, it is also a relatively low-value, low-efficiency and highly subsidized water user, and it is expected that with increasing water scarcity the agricultural sector has to produce more food and fibre in the future using less water than it does today.[6] The present trend among decision-makers is to increase the resources devoted to the rehabilitation and modernization of irrigation systems in an attempt to raise both cropping

---

[1] Food and Agriculture Organization (FAO) 1990, 1993
[2] The World Bank/ICID Dec. 1989b, p.16
    Average irrigation lending per year was US$ 37 million in the 1950s, US$ 343 million in the 1960s, US$ 1,120 million in the 1970s, US$ 1,273 million in the 1980s, and US$ 1,032 million so far in the 1990s. (in 1991 US Dollars; WB 1994, iv)
[3] See FAO 1990, p.13. It is assumed by Carruthers and Clark that they range from US$ 1,500/ha in Latin America and the Far East to nearly US$ 2,500/ha in the Near East and Africa. (1981, p.134) In 1984, in Indonesia, for example, the capital costs/ha average US$ 4,850; in the Philippines they increased from US$ 1,534 to 2,876 after 1980. (see Small and Adriano 1989, p.37; Small and Adriano 1989, pp.193-194)
[4] See The World Bank 1994, p.121
[5] See van Tuijl undated, p.2
[6] FAO 1994, p.10

intensities and crop yields from the existing capital investments.[7] These are extremely low-cost investments as a consequence of the large element of sunk costs, i.e., once they are paid, these costs are labelled as 'sunk', and the costs per hectare may be only one-third to one-fifth of those of new projects.[8]

A large body of literature evaluates the performance of public irrigation systems, and the most common conclusion is that the large formerly constructed and operated public irrigation schemes, which were promoted with high priority and massive public resources, failed or were less successful than expected. In 1977, the United Nations Water Conference estimated that by 1990 approximately 92 million ha would require improvement.[9] The World Bank and the International Commission on Irrigation and Drainage assume that between 50 and 70% of the irrigation systems of the world are in need of rehabilitation, and that almost all networks call for some degree of modernization.[10] The FAO considers that, currently, the overall performance of many irrigation projects is much poorer than expected.[11] Inadequate operation and maintenance and inefficient management have contributed to many socio-economic and environmental problems. The World Resources Institute confirms that „important performance measures, such as acreage irrigated, yield increase, and efficiency in water use, are typically less than projected when investments were made, less than reasonably achievable, and less than attained by private irrigators who operate more controllable decentralized systems".[12] The International Food Policy Research Institute emphasizes that „dissatisfaction with the performance of irrigation projects in developing countries is widespread. Despite their promise as engines of agricultural growth, irrigation projects typically perform far below their potential".[13]

Indicators of project failures are manifold and can be summarized as follows:[14]

1. Cost and time overruns.
   The implementation of the projects usually takes longer than anticipated by the official planners. Cost and time overruns occur with depressing regularity.[15]
2. Faulty design and lack of drainage system.
   Lack of a drainage system, or its insufficient technical layout, and insufficient control structures are observed in many projects all over the world.[16]

---

[7] The World Bank/ICID Dec. 1989b, p.17
[8] Carruthers and Clark 1981, pp.134-137
[9] United Nations Water Conference 1977; Biswas 1985
[10] The World Bank/ICID Dec. 1989b
[11] FAO 1990
[12] Repetto 1986, p.3
[13] Small and Svendson 1992, p.1
[14] See Bottrall 1981; Carruthers and Clark 1981; The World Bank 1983, 1994; Hotes 1983; Biswas 1985; Repetto 1986; Barrow 1987; Small et al. 1989; BMZ 1990; Small and Carruthers 1991; Carruthers 1993; van Tuijl undated; Turral 1995
[15] A review of more than 200 World Bank funded irrigation projects came to quite surprising results: time overrun is not as serious as believed, and the average implementation delay, 1.7 years, was only slightly above the average for all projects. (The World Bank 1994)

3. Overestimation of the anticipated rate of increase in output (or increase in yields).
4. Capital cost and recurrent cost recovery.
   Cost recovery has fallen short, and the operation and maintenance (O&M) of the projects need a steady flow of subsidies.
5. Poor operation.
   The operation of irrigation projects shows low water-use efficiencies and low water-conveyance efficiencies, resulting in inadequate and inequitable water distribution.[17]
6. Poor maintenance.
   The maintenance of the delivery and drainage infrastructure is inadequate due to chronic understaffing and low budgets, and maintenance activities are deferred, reducing the channels' capacity with negative impacts on the operation and the environment.
7. Misallocation and wasteful use of water.
   The systems often have a low water conveyance efficiency and a low farm efficiency due to poor operation and maintenance, inadequate farming practices and the adoption of ineffective irrigation technologies. Land levelling and water management at the farm level are poor.[18]
8. Harmful environmental impacts.
   Negative environmental impacts such as high groundwater levels, waterlogging and salinity, water-borne diseases, sedimentation, and water contamination have been extensive.

As Small and Carruthers point out, some of these problems are interlinked, and „no single reason, not even financial problems, can be put forward to explain failure of irrigation investments to realize their maximum potential".[19]

Although mentioned last, the negative environmental impacts of public irrigation projects impose high costs, and the seriously widespread environmental problem of high groundwater levels, waterlogging and salinity which, for a long time, has attracted great attention, is selected for study. The installation of an irrigation system converting agriculture from rainfed to irrigated leads inevitably to a major change in the local environment with impacts on the natural and social environ-

---

[16] The World Bank report (1994) states that drainage has been an explicit element of more projects than any other physical feature; it is also prominent in legal convenants.

[17] Water-use efficiency, or overall efficiency, is the ratio between the quantity of water placed in the root zone and the total quantity supplied to the irrigated area. This indicator represents the efficiency of the entire operation between diversion, or source of flow, and the root zone. Water-conveyance efficiency of the main, lateral and sublateral canals to the farm outlet is defined as the volume of water delivered in relation to the total quantity of water supplied to the area. (Bos and Nugteren 1974, pp.9-11; see DVWK 1983)

[18] In the developing countries, as much as 60 % of the water diverted or pumped for irrigation is assumed to be wasted. (FAO 1994, p.10) Farm efficiency is defined as the ratio between the quantity of water placed in the root-zone and the total quantity under the farmer's control. (Bos and Nugteren 1974, pp.9-11)

[19] Small and Carruthers 1991, p.3

ment, and vice versa.[20] Abernethy and Kijne identified four main categories of
environmental interactions: (a) interactions which occur within the domain of the
irrigation projects, e.g., high groundwater levels, waterlogging and salinity; (b)
those which occur outside the project area and affect it, e.g., upstream land use; (c)
those which are caused within the project area by its usage, and which cause environmental impacts and costs outside, e.g., water contamination; and (d) those
which occur within the agricultural project, and do not directly reduce its productive capacity, but have other negative consequences, for example, water-borne
diseases.[21]

The most common causes for major concern are the extremely serious problems
of high groundwater levels, waterlogging and the irrigation-induced salinization of
soils which affect the productive capacity of the systems themselves and, thus, the
overall target of irrigated agriculture, the increase of yields. High groundwater
levels and waterlogging have the most detrimental effect on crop yields, either
directly, or in association with soil salinity. The USAID estimates that „to a degree, almost every irrigation system in the world must contend with prospects of
waterlogging or salinization".[22] Of the approximately 270 million ha that are irrigated worldwide 20 to 30 million ha are severely affected by salinity, with another
60-80 million ha affected to some extent.[23] Other figures reveal that soil salinity is
seriously affecting the productivity of one-third of the world's irrigated land, and
the World Bank assumes that „waterlogging and salinity are encountered at a significant level in about 15 million ha of irrigated land in the arid and semi-arid
zones in developing countries".[24] In Pakistan as many as 11 million ha of the 15
million ha of irrigated land are already suffering from salinization and waterlogging. Afghanistan, Egypt, Iran, Iraq, Sudan, Syria and the former USSR also have
similar problems.[25] Pakistan, for example, devoted most of its efforts to improving
drainage on existing irrigated land to avoid losses to salt and waterlogging.[26] India
and the ex-Soviet republics of Central Asia are struggling to maintain production
in the face of salt and waterlogging which has resulted from delayed drainage
investments.[27] In Xinjiang, northwest China, 1.8 million ha of the 3.4 million ha of
arable land have had to be abandoned due to salinization.[28]

For many decades, the general physical causes and the principal techniques for
preventing and dealing with high groundwater levels, waterlogging and salinity
have been well known and are well established.[29] The major causes are excess
water input into the project areas, poor maintenance of the delivery and drainage

[20] For details see Baumann et al. 1984; Goldsmith and Hildyard 1984; Petermann 1993a; GTZ 1993; FAO 1994
[21] Abernethy and Kijne 1993, pp. 75
[22] USAID 1981, p.15
[23] FAO 1990; see also Postel 1989; Biswas 1985; Dhillon et al. 1981
[24] The World Bank 1989, p.50
[25] Ahmad and Singh 1991, p.298; see Barghouti and Le Moigne 1991; El-Ashry 1980
[26] For the Indus valley in Pakistan, see The World Bank 1994, p.32
[27] The World Bank 1994, pp.4-5
[28] See Betke 1989
[29] See FAO/UNESCO 1973; Diestel 1982, 1987

channels, and the inadequate design or total lack of drainage systems.[30] To control high groundwater levels and salinization, a high water conveyance efficiency with little water loss is desired. An effective natural or man-made drainage system is needed for the entire drainage area including the fields, but percolation should not be reduced to zero to allow for the leaching of accumulated salt. High water applications at on-farm level exceeding crop evapotranspiration requirements should also be controlled. The major control means are:

* investments in adequate drainage systems, and their maintenance;
* investments in delivery systems with low percolation coefficients, and their maintenance, and
* adequate water inputs for irrigation as well as for leaching, regulated by economic incentives and/or by the administration of water distribution. (see Table I.1.)

The International Action Programme on Water and Sustainable Agriculture Development concedes that solutions are generally known, and that the problem can be controlled „*if* appropriate measures can be implemented",[31] and Diestel estimates that „simply because such *common-sense* measures are not applied (...), every day the definite salinization of some 100 ha or so is witnessed in this world".[32]

Table I.1. Factors leading towards high groundwater levels and salinization, and technical, managerial, and economic remedies

Excess water input at on-project level
→ investments in and maintenance/repair of the delivery channels
→ improving main system management

Excess water input at on-farm level
→ reduction of the applied amount of water through economic incentives and/or
→ the operation of the main system

Drainage of excess water
→ investments in and maintenance of the drainage channels

However, irrigated agriculture struggles with these problems, and since the early 1980s international donor organizations and non-governmental organizations have concentrated considerable efforts and money towards the rehabilitation and modernization of large-scale irrigation schemes. An International Action Programme on Water and Sustainable Agriculture Development (IAP-WASAD) has been formulated by the FAO to be implemented in the 1990s with identified priority areas of action including, beside other issues, „efficient water use at farm level;

---

[30] For details see GTZ 1993; The World Bank/ICID Dec. 1989a and 1989b
[31] FAO 1990, pp.15-16 (in the original without italics); see Carruthers 1993
[32] Diestel 1993, p.189 (in the original without italics)

waterlogging, salinity and drainage".[33] The World Bank started an intensive research and training programme on the same issue in 1989, with the technical assistance of the International Commission on Irrigation and Drainage.[34]

Different disciplines and professions are dealing with this widespread and prevalent environmental problem; each profession has its own solutions and its neglected opportunities, and their various approaches stress different aspects. From an engineering-oriented viewpoint, hardware solutions are suggested which Chambers paraphrases as the 'plug, pump and drain' solution of engineering science.[35] In the 1980s it became conventional knowledge that to improve performance, irrigation systems have to be examined in their entirety, with a whole system approach.[36] This approach focuses on management problems and managerial changes, or, as Carruthers emphasized, that „the main thrust of the new strategy must now be to address questions of management", after „irrigation technology has been a victim of poor planning, construction and management".[37] Chambers also centrally addresses the 'main system management' in South Asian irrigation schemes which, in itself, could generate high groundwater levels, waterlogging and salinity, apart from other problems.[38] 'Management' has become a broad term for very different views and approaches, but they have one subject in common: they focus on management units and their operation and maintenance activities.[39] Conventional economists operate in a different dimension, with prices and incentives. Because water is an increasingly scarce and valuable resource with finite supply options, supply-driven solutions, which dominated in the past, should be abandoned. Demand management should re-allocate the water between different users, e.g., within the agricultural sector, or from agriculture to other higher-value uses.[40] Water-use efficiency, as an economic target, is regarded as attainable whenever the pricing method affects the demand for irrigation water. If the farmers, as rational optimizers, use too much water, they can be discouraged by having to pay more for it. Achieving water-use efficiency is, at the same time, regarded as a means to deal with high groundwater levels, waterlogging and salinization.

However, the important question remains as to which remedies can 'effectively' and 'adequately' deal with high groundwater levels and salinization, and why appropriate remedies are not implemented. Why are decisions on technical, managerial and/or financial means neglected, and why are such decisions made by those

---

[33] FAO 1990, p.5

[34] The World Bank/ICID Dec. 1989b

[35] He treats ironically the recommendations of irrigation engineering which comprise of: „(i) reduce inflow through lining ('plug'); (ii) remove groundwater through pumping ('pump'); (iii) remove surface and groundwater through drainage ('drain'); (iv) educate farmers in water management" (1988, p.76). Chambers criticizes the concentration on hardware remedies when they are seen as the only remedies.

[36] Irrigation systems are viewed as socio-technical open systems embedded in a socio-economic, natural and an institutional environment. (see Huppert and Walker 1988)

[37] Carruthers 1993, p.17

[38] See Chambers 1988: Main System Management: The Central Gap, pp.105-132

[39] The fragmentation of responsibilities among the construction, operating and agricultural agencies is regarded to be part of the management problem.

[40] See FAO 1994

concerned with results other than the environmentally desired outcomes? What are the constraints facing the participants involved in large-scale public irrigation systems where state agencies are the management units? In particular, why do the negatively affected users of the schemes, i.e., the farmers, take no action towards halting and reversing the process, when, in principle, they control some means for doing so? It is assumed here that the causes which generate the negative environmental effects lie in the particular institutional setting which does not provide adequate incentives for the multiple participants involved.

Theoretical Approach

The economic study of non-market decision-making addresses the provision and supply of goods and services by the participants of the political and bureaucratic systems, and scrutinizes, particularly, the central role of state bureaucracies. Anthony Downs' economic theory of democracy (1957) and bureaucracy (1967) and William A. Niskanen's theory of bureaucracy (1971) are crucial to this approach. Anthony Downs focused on how bureaucracies perform their social functions. He challenged Max Weber's notion that bureaucracies are passive respondants to government decisions by making the fundamental assumption that bureaucrats and politicians, like all individuals, are self-interested utility maximizers whose rational choices are limited by an individual's capacity for information processing. A bureaucracy's formalized internal structure would generate distortions and asymmetries in information; bureaus would define problems and solutions within their areas of responsibility, and, therefore, neglect the connectability of interrelated issues; bureaus would secure autonomy and the ability for performance by avoiding conflicts with other bureaus and with their political environment. Downs, however, assumed that the self-interest of politicians and the competition amongst political parties would guarantee that the citizens' preferences are transferred into the government's demand on the bureaucracy's supply, and that social institutions determine whether the outcomes are efficient. William N. Niskanen was interested in the endogenous motives and dynamics of bureaucracies' growth. He regarded bureaucrats as 'budget maximizers', because they can serve their utility function best with large budgets. This would be the main force behind exploding public expenditures. In his formalized model of bureaucratic supply, the bureau has overwhelming power over the government/sponsor because of information asymmetries. He presumes that the bureau would have the same type of market power as a monopoly that presents the market with an all-or-nothing choice. This would result in producing outcomes exceeding those demanded by the government/sponsor and the citizenry. Niskanen's model is valid for restrictive conditions: it comprises a bilateral relationship between one bureau and a government, and does not seek validity if more than one bureau is considered, or if inter- and inner-bureaucratic relationships become important. The process on how, and if, demands are generated through political mechanisms was outside his inquiry, and his theory did not intend to provide an explanation for bureaucratic performance.

The object of this study calls for an approach which centrally considers that a multiplicity of political, bureaucratic, civil and economic decision-makers are involved, who decide on the provision and supply of irrigation services, on their financing, and on the use of the resource, all of them having differing and conflicting preferences, generating collective action problems. The analysis essentially concentrates on the internal conditions of bureaucratic performance, and it discusses, particularly, how state bureaucracies deal with externalities, because the installation of an irrigation system, i.e., a common-pool resource system, introduces two types of physical linkages among all formerly independent farmers receiving their water from one supply system within a drainage area. The interdependencies refer to the joint use of a water-delivery system and to the joint use of a drainage system. The first interdependency generates provision and maintenance problems for the delivery system, and problems of the allocation of water where interfarmer conflicts arise due to the substractability of resource units. The use of water creates the necessity for the provision and maintenance of drainage infrastructures to discharge effluents from private farm-firms, which, at the same time, prevent high groundwater levels and salinization. However, the important question is how collective action among individual users is stimulated, and how both interdependencies, which cause externalities, are mediated by rules for the provision, maintenance and water allocation to limit individualistic strategies with inefficient outcomes. As the peculiar nature of irrigation and drainage systems requires the establishment of rules, a problem of major concern is how the state and the management unit can enforce them to ascertain collective action among interdependent users. This is particularly important in the case of infrastructure-intensive resources such as large-scale irrigation systems; in the case where many transformation and transaction activities are necessary to achieve an adequate, reliable and equitable water supply and to sustain the irrigation and drainage infrastructure, and in cases with a large number of polluters and victims.

Elinor Ostrom, a scholar of the New Institutional Economics (NIE),[41] provides a general theoretical framework for analyzing institutional arrangements such as the 'market', the 'political' and the 'bureaucratic' system, and the 'bargaining of individuals or groups of individuals', and for the selection of institutions. This approach seeks to explain characteristics of social outcomes on the basis not only of agent preferences and optimizing behaviour, but also on the basis of institutional features. Institutions are characterized by rules-in-use which shape the individuals' behaviour through the impact of their incentives. Rules-in-use such as, for example, boundary rules, authority rules, etc., form the structure of action situations with multiple participants being involved, and the rules offer explanations of actions and results. Scholars who depart from the NIE have concentrated theoretical efforts on studying common-pool resource systems, on the creation of design principles for long-enduring self-governed irrigated common property regimes, and on the analysis of public infrastructure policies by drawing insights from numerous empirical studies and various theories, e.g., the economic theory of democracy and bureaucracy, the theory of industrial organizations, the theory of property

---

[41] Some scholars use the term 'New Institutionalism' as the general banner under which different concerns with institutional features have been elaborated. The Institutional Rational Choice approach (IRC) shares central issues with the NIE but centres on institution selection and on their design. (see Shepsle 1989)

rights and transaction costs, and theories on collective action.[42] Table I.2. shows the relation between the technical, managerial and economic means for groundwater level and salinity control to be decided on by those responsible, and the associated theories to explain different aspects of the setting.

Literature Review and Open Questions

Recent criticism of state-regulated irrigation schemes and their *bureaucratic management* is that irrigation agencies are monopolies, and that they face incentives neither for sustaining the infrastructure facilities nor for efficient water use. They receive their budget from central budget allocations, and the decisions made on the O&M budget are not related to the performance of a project, but to political considerations. New investments are favoured over the completion of older schemes, and the infrastructure facilities are inadequately maintained; investments in drainage systems and their maintenance are neglected or poor. In many countries, the beneficiaries must finance the recurrent costs for operation and maintenance, and in some countries they have to, at least partially, repay the initial costs; but water charges are not assessed to fully recover recurrent costs, and the water charges collection levels are very low. The allocated O&M budgets, however, show no relation to the water charges. Water charges do not instigate water-use efficiency, because they are flat rates, unrelated to the applied volume of water, and the opportunity and social costs are not accounted for. Water supply scheduling, which is to guarantee an equitable, reliable and adequate water supply, is poorly enforced by public officials, and is more often than not opposed by the irrigators; tail-end irrigators are deprived of water supplies and water is applied in excess. Public officials receive low salaries, and diligent service and good performance are not rewarded; public officials are plagued by shirking, and in some countries by corruption. Governmental control over the bureaucracy's services is weak, and the consumers/beneficiaries have influence on governmental decisions only through the mechanisms of the political system. Public officials are not accountable to them, and they are in no way dependent on those they serve, although irrigation is a service with customers and users.

There is a growing advocacy for *water markets*, which are regarded as an adequate means for achieving economic water-use efficiency, i.e., the allocation of resources in ways that maximize their contribution to human well-being.[43] Water markets would also solve salinization.

---

[42] See Ostrom 1990; 1992b; Ostrom et al. 1993, 1994
[43] See, for example, Rosegrant and Binswanger (1994), who consider the establishment of tradeable property rights in water and the development of markets in water rights as a policy option in response to poor economic performance.

Table I.2. Theories on decision-making in large-scale public irrigation systems

Items	Decision-making units	Theories
Investments in irrigation infrastructure * funds * planning and construction * financing	Legislature, government, ministries	Theories on non-market decision-making * Economic theory of democracy (Downs) * Economic theory of bureaucracy (Tullock, Downs) * Economic theory of bureaucratic supply (Niskanen)
Executing investments	State agencies; private sector	General theories on institutions * Transaction cost theory (Coase, O. Williamson) * Property-rights approach (Alchian, Demsetz, Coase, Bromley)
Operation and Maintenance * funds * financing * supply of services	Government, ministries Government, Ministry of Finance State agency, farmer groups	* Theory of the firm and contracts (Coase, Tirole) * Principal-Agent approach (O. Williamson, Jensen and Meckling) * Institutional Rational Choice (E. Ostrom, Shepsle)
Use of common-pool water resource system Use of collective good 'drainage system'	Irrigators	Theories on the commons * Commons dilemma (G. Hardin) * Common-property regimes (Ciriacy-Wantrup, E.Ostrom) Theories of collective action (M.Olson, R. Hardin) Theory on interest groups (M. Olson) Theory on rent-seeking (Krueger, Repetto) Cooperative and non-cooperative game theories
Use of water Irrigation and farming practices Investments at on-farm level	Individual farmers	Private farm-firms as profit optimizers

However, the internalization of externalities through markets needs clearly defined property rights, and also perfect and competitive markets and zero transaction costs to yield efficient outcomes. The nature of irrigation and drainage systems, however, creates numerous sources of market failure, and transaction costs are high. Only a few countries have well-functioning water markets, and empirical research on water markets with regard to their ability for controlling salinization has remained poor.

Research on *irrigated common-property regimes*, where farmer groups own, operate and maintain the systems at their own costs and with their own rules, be they small-scale gravity or groundwater irrigation projects, or local subsystems within large and complex gravity schemes, has shown that these institutional arrangements are viable alternatives to state regulation and markets. Empirical and theoretical research on small-scale farmer-owned and -managed schemes provides explanation on water-use efficiency, financing investments and financing operation and maintenance, and it is concluded that common-property regimes are, under certain conditions, able to efficiently provide and manage common-pool resource systems, to hold officials accountable to the users, and to limit free-riding. These regimes provide effective control mechanisms, with low transaction costs. No empirical studies on high groundwater levels and salinity problems are available for farmer-owned and -managed systems, but it is evident that these schemes show some advantages if compared with state-administered large-scale systems.

Jointly managed public-farmer irrigation systems, where irrigator groups operate and maintain the lower levels of the schemes, show a great variety with regard to their institutional setting. The main differences concern the state bureaucracy, (e.g., whether it is financially dependent or semi-autonomous,) and the design of the participating farmers' groups, (e.g., their range of decision-making, the incentives for participation, their internal structure).[44] It has been recognized that water distribution and the maintenance of the delivery systems has improved, and that the contributions of the beneficiaries towards O&M costs have increased. The importance of restricting government subsidies has prompted more than 20 countries to promote programmes of irrigation management transfer. Some of the most significant changes affecting irrigated agriculture now taking place in the developing countries relate to irrigation finance policies, to the basic structure of irrigation agencies, and to the relationship between irrigation agencies and higher-level government authorities, on the one hand, and organized groups of farmers, on the other, to ensure the long-term and sustainable productive use of natural resources.[45] Of foremost importance is how the performance of irrigation bureaucracies as the managing units can be improved, and how they can foster conscientiousness among staff to make the organization's goals the basis for their own actions.[46] This

---

[44] There is a wide range of options for joint management, and design principles for viable public-farmer institutions remain an important research issue, e.g., what types of interaction and external assistance are required for long-term viability, and how is cooperation institutionalized. (see Subramanian, Jogannathan and Meinzen-Dick 1995)

[45] Small and Svendson 1992, p.24; The World Bank 1993

[46] The World Bank 1994b, pp.94-95

relates to the issue of what changes in the relationship 'bureaucracy-consumers' would improve the efficiency of investments, operation and maintenance. The conditions under which jointly managed public-farmer systems are able to success-fully alleviate high groundwater levels, waterlogging and salinization are open to question.

As the institutional conditions which promote efficient and equitable water allo-cation are not identical with the conditions for effective groundwater and saliniza-tion control, there is a need for empirical and theoretical research on the limits and potentials of the varying institutional arrangements to deal with this issue. This study is the first empirical research which analyzes high groundwater levels and salinization in large-scale public irrigation projects from an institutional perspec-tive.

Research Hypotheses and Goal of the Study

The general assumption of this study is that high groundwater levels, waterlogging and salinization are the outcome of the rational decisions of all participants in-volved, i.e., the political, bureaucratic and economic decision-makers. If a public irrigation scheme is poorly managed and maintained, if resources are inefficiently used by individual economic agents, and if high groundwater levels and salinized soils prevail, then this is interpreted as the result of the rational behaviour of the multiple participants within uncertain and evolving circumstances, and within the given institutional setting. From empirical research which departs from the New Institutional Economics, it is presumed in this study that the rules-in-use provide incentives for an efficient or inefficient allocation of resources; for the sustainment of public infrastructure investments, or for neglected maintenance; for proper management, or poorly performed systems; for the efficient use of natural re-sources, or for their overuse, and rules-in-use may, or may not, internalize negative externalities.

The study aims to reveal whether the New Institutional Economics also pro-vides an explanation for the prevalence of the selected environmental problem in a large-scale public irrigation project. The central research hypothesis is that, in a highly centralized non-market decision-making setting, institutional constraints impede the implementation of adequate groundwater and salinity control means. The case study aims to find out whether the set of rules in two joint management models provide functional/dysfunctional incentives contributing, or not contribut-ing, to high groundwater levels, waterlogging and salinity; it discusses, foremost, why water saving and drainage as the central groundwater and salinity control means are not forthcoming in centralized decision-making, and whether a higher accountability of the management units to the users improves the chances for the implementation of salinization control means.

Analyzing public irrigation systems with regard to the institutional conditions which contribute to the environmental problem, independent but interlinked action arenas are considered: the planning and design process; the provision of goods and services comprising decisions on the kind of services, the quantity, the quality, how to finance, how to monitor, etc.; the production, e.g., the execution of investments and the rendering of O&M services including groundwater and salinity control where transformation and transaction activities are costly; the joint use of the irrigation and drainage infrastructure and the substraction of resource units where the inducement of collective action is important; the use of water units as input factors in farm economy with agricultural effluents being produced.

The Case Study

The large-scale Lower Seyhan Irrigation Project in Turkey was selected for study. From its beginning in the 1960s until 1994, the project showed characteristics of a highly centralized non-market decision-making setting: dependent bureaucracies executed investments relying completely on centrally allocated budgets, and a state agency operated and maintained the scheme together with Water User Groups. This setting recently underwent changes, and since 1994, a different concept of joint management has been introduced where Water User Associations have received greater autonomy.

The Lower Seyhan Irrigation Project was designed to divert irrigation water to 170,000 ha of arable land, the fields of more than 20,000 privately owned farms. The installation of the irrigation project would promote higher yields and revenues, a diversity of higher-value crops, and cropping throughout the year; but the realization of the project's objectives was impeded by high groundwater levels and saline and alkaline soils. At the end of the 1980s, almost the entire service area was in need of rehabilitation work.

The study analyzes the project's implementation process, the operation and maintenance of the irrigation scheme, and the use of the resource in farm economy whilst focusing on the institutional conditions which contribute to the prevailing and widespread environmental problem. The study attempts to trace the decision-making of the multiple decision-makers over a time period of more than 30 years. The analysis of the project's implementation process is based on planning documents, feasibility studies, evaluation reports and on various data obtained from the international funding and the Turkish agencies. It uses results of research work conducted by Turkish scientists and state agencies to reveal the development of high groundwater levels and salinity. The performance of O&M, the state agency's organizational structure, and its cooperation with the Water User Groups and the Water User Associations is evaluated on the basis of documents, insight views, and interviews/personal communications with DSI staff, leaders of Water User Groups and Water User Associations, ditch-riders, and farm operators. Literature from political scientists on Turkish bureaucracy and on interest groups in Turkey has been used as a background for the interpretation of empirical observations. Empirical on-site research was undertaken during several long study visits in Turkey, the first in 1988, a long stay in 1989, again in 1991, and the last in June 1995. The visits concentrated on revealing operation and maintenance activities carried out by the state agency, the Water User Groups' and Water User Associations' contributions to O&M, and, finally, the economic impact of high groundwater levels and salinity on farm economy. This last point caused some difficulties, as there are no on-field studies available which empirically evaluate the negative impacts on farms.

Outline of the Study

Chapter One reviews the economic theory of non-market decision-making, and its view of the central role of bureaucracy. It starts with the Downs' and Niskanen's models, focusing on the latter which is still important in economic science. The chapter then summarizes existing theoretical expositions on the nature of irrigation and drainage systems with which decision-making units have to deal. Finally, it displays the extended method of institutional analysis developed by Elinor Ostrom.

Chapter Two surveys a series of empirical and theoretical research on large-scale public systems and small-scale common property regimes, concentrating on the institutional conditions which impede, or promote, efficient water allocation and the sustainable maintenance of the irrigation networks. Discussed are whether jointly managed public-farmer irrigation systems show potentials for groundwater and salinity control, and the way in which the management units try to internalize negative externalities.

Chapters Three to Five comprise the case study. They analyze the first public-farmer setting with its highly centralized decision-making structure and its outcomes, and then discuss the potential for controlling groundwater and salinization in the second public-farmer management structure.

Chapter Three, after describing the natural conditions in the area prior to the project, analyzes the implementation process of the three subprojects carried out by two state agencies, with particular reference to high groundwater levels, water-logging and salinity control. The shortcomings in relation to the investment and implementation strategies are ascertained by reviewing the evaluation reports of the World Bank and the Turkish state agencies. It then evaluates the effects of the subprojects and of the rehabilitation project on groundwater and soil conditions.

Chapter Four focuses on the contributions to high groundwater levels, water-logging and salinity by the operation and maintenance of the jointly operated scheme with its two joint management settings. Its main objective is to reveal the reasons why negative external effects are produced, and why they are not prevented.

Chapter Five analyzes the positive impacts of the project on yields and net incomes, and the negative effects of high groundwater levels and salinized soils. It evaluates the costs deriving from high groundwater levels and salinity, and the costs incurred by the farmers for control means in cotton farming under the two public-farmer settings, and it discusses the difficulties of collective action.

The final conclusions in Chapter Six combine the results obtained from the case study with prominent hypotheses from empirical research, and discuss their relation to the general assumptions and working hypotheses of the New Institutional Economics. It finally presents hypotheses for an effective strategy to control groundwater and salinization, and necessary research therein.

Chapter One
Theories on the Provision and Supply of Goods and Services Through the Political and Bureaucratic System

In most developing countries, the development of water resources has been dominated by the state, and the planning, construction, operation and maintenance of large-scale irrigation projects has been subject to state bureaucracies. The provision of government financing developed from a historical situation after World War II, when no private investors were able, or willing, to finance these large and costly infrastructures because their scope was beyond private endeavour. Governments thus stimulated economic development through infrastructure financing. The public provision of infrastructure financing was, and still is, regarded as a necessity, because under perfect competition - according to the standard economic conclusion - underinvestment would occur, in that individual firms could not appropriate the entire economic benefit.[1] In its policy paper referring to water resources development, the World Bank points to the overwhelming trend towards government provision until the early 1980s, because „it was believed that only the state was able to handle the large investments and operations necessary for irrigation and water supply systems and that the crucial role played by water justified government control".[2] Government financing arrangements would be best able to limit free-riding, realize economies of scale in production, make the best use of technical expertise, and government intervention would internalize social costs when the number of individuals is large. The fiscal crisis in the developing world that began in the early 1980s, however, has demonstrated the weakness of governmental interventions, and experiences in public irrigation systems have not validated the positive assumptions. The misallocation of resources, the poor performance of water resource systems, the deterioriation of public infrastructure facilities and the prevalence of negative environmental effects have exposed the serious institutional deficiencies of many goverment agencies responsible for water resources, and have put to question, in general, the governments' role. The literature has thus become dominated by the search for substitutive institutions, e.g., the market and/or civil organizations, which could offer what the states were not able to provide.

A matter of major concern in this chapter is whether the various approaches of the New Institutional Economics provide explanations for the objective of this study, i.e., the control of salinization in a non-market decision-making setting. While the classical political economists have concentrated on the origins and the

---

[1] See de Horter and Zilberman 1990; Small and Carruthers 1991, pp.25-26
[2] The World Bank 1993, p.100

functions of the state, and its fiscal policy, early neo-classical economics has treated the state and governments as a 'black box', lying beyond economic considerations. Early economic theorists of democracy and bureaucracy (Downs, Niskanen) who applied the economics methodology to politics, recognized the importance of political decision-making and political institutions on economics, and they provided relevant theoretical assumptions concerning political and bureaucratic decision-making. They questioned, however, how an efficient allocation of resources is affected by the decision-making of political and bureaucratic agents, but first, the theories and approaches labelled with the term New Institutional Economics (NIE) theoretically and systematically reflected the central role of institutions for the individuals' decision-making. Elinor Ostrom, a scholar of the NIE, developed a general method for institutional analysis where decision-making is understood as the outcome of strategic games between multiple participants, all of whom have preferences, well specified objectives, and are incentive-constrained. She provides a framework for analyzing public irrigation infrastructure policies and bureaucratic water resources management considering transaction and information costs, on the one hand, whilst on the other hand she analyzes collective action and collective outcomes, and the elimination of the free-rider problem not only in terms of individual motivation but on the basis of institutional features. In this study, her approach is applied to identify the institutional constraints and the incentives facing the participants, which could explain the prevalence of high groundwater levels and salinization in a large-scale public irrigation project.

1.
Economic Theories of Democracy and Bureaucracy

The economist's interest in the provision and supply of goods and services through political and bureaucratic mechanisms lies in how these institutional arrangements amalgamate the individuals' preferences to generate collective results, and how collective decisions respond to scarcities, tastes, preferences and opportunities. The economic study of non-market decision-making is crucial to the Public Choice approach which has scrutinized the central role of politicians and bureaucrats, and the mechanisms to generate collective results, because the public provision of goods and services through non-market mechanisms depends upon decisions made by the participants in the political and bureaucratic system.[3]

---

[3] The Public Choice approach is defined „as the economic study of non-market decision-making, or simply the application of economics to political science". It makes „the same behavioral assumptions as general economics (rational, utilitarian individuals)"; it „often depicts the preference revelation process as analogous to the market (voters engage in exchange, individuals reveal their demand schedules via voting, citizens exit and enter clubs)", and it asks „the same questions as traditional price theory (Do equilibria exist? Are they stable? Pareto-efficient? How are they obtained?)" (Mueller 1989, p.3).

The basic behavioral postulate of the Public Choice approach, and that of economic theory in general, is that politicians and bureaucrats, like all individuals, are intentionally and rationally bound, self-interested utility maximizers.[4] In An Economic Theory of Democracy, Anthony Downs called attention to those who might be called on to provide public goods and services demanded by the voters/ taxpayers.[5] He focused on the rational behaviour of politicians whose utility function is not seen to be consistent with the public interest(s), or as Frey paraphrased it, politicians do not pursue the social wellbeing, but their own interest. First and foremost they want to secure their own reelection.[6] Downs assumed that the competition among different parties makes sure that the voters' wishes are considered in the political process. Recent research makes the point that although the individuals' preferences are amalgamated to generate collective results through collective-choice mechanisms, i.e., voting and delegation, the outcomes of elections do not necessarily provide a clear indication of majority preferences as to which type and amount of goods should be provided and how funds should be allocated.[7] The difficulties in expressing preferences through voting are that the citizens' diverse preferences are not automatically translated by voting mechanisms into a well-defined order for a variety of goods for a community as a whole. Outcomes are, in fact, strongly affected by the presentation order of alternatives and other aspects of voting. Voting decisions are rarely confined to provision decisions concerning only one good, even if voting mechanisms adequately translate individual preferences for single goods. The officials for whom citizens usually vote are responsible for decisions on the provision of many goods and services. An official may, therefore, closely represent a citizen when regarding one type of infrastructure, but not another. The intense preferences of some voters cannot influence the voting, as all votes carry equal weight. A situation, however, where the trading of votes were possible could be beneficial when there are voters with strong preferences and voters who feel indifferent. Should voters lack a sense of responsibilty for their choices, they may show little interest, and thus not enquire further about the issues. Those more motivated to advocate the provision of a particular type of good through interest groups and other political activities are usually those who will benefit more than others from its provision. It may be more costly, in time and energy, to spread the costs of provision evenly over a population opposing the provision of a good that benefits one group more than others than bearing the added costs of taxation. Effectively organized groups may be able to generate an overinvestment and thus a disproportional benefit for themselves by mobilizing political support. Groups, however, that are not effectively organized may not be able to affect enough electoral support in national or provincial elections to gen-

---

[4] See Downs 1968, pp.4; Tietzel 1981, pp.115; Schäfer and Ott 1986, pp.46
    The extreme rational choice model makes unrealistic assumptions about the information-processing capabilities of individuals, and about the evaluation and calculation processes, therefore the model was changed into what is now called bounded, or limited, rationality. (see Herbert A.Simon 1957)
[5] Downs 1957
[6] Frey 1991, p.492
[7] See Arrow 1951; Buchanan and Tullock 1962; Olson 1965, 1982

erate investments that would lead to a situation of economic benefits higher than costs.[8]

While the decisions on goods and services to be provided by the government are left to collective-choice mechanisms, the governmental bureaus, i.e., state bureaucracies, usually execute investments and supply services. Theories of bureaucracy in sociology and political sciences tended to be dominated by Weberian notions of impartial, efficient service by the public officials concerned to serve the public interests as interpreted by their elected government.[9] Early challengers started to develop a theory of administrative hierarchies analogous to the 'theory of market', subjecting bureaucratic agents to critical scrutiny and noting the inevitable loss of control associated with this form of economic organization.[10] Anthony Downs' theory centrally addresses the factors which significantly affect a bureau's decision-making in performing its social functions. Like Gordon Tullock, he makes the fundamental premise, that „bureaucratic officials, like all other agents in society, are significantly - though not solely - motivated by their own self-interests",[11] and he regards bureaucracy no longer as a passive respondant to government decisions. The 'utility function' of the public officials in Downs' theory is constituted of self-interested and altruistic goals, and „whether or not the public interest will in fact be served depends upon how efficiently social institutions are designed to achieve that purpose. Society cannot insure that it will be served merely by assigning someone to serve it".[12] His theory on bureaucratic decision-making rests on three central hypotheses: firstly, that bureaucrats „seek to attain their goals rationally (...) they act in the most efficient manner possible given their limited capabilities and the costs of information," and that some uncertainty remains in decision-making. Secondly, that the bureaucrat's utility function is a complex set of goals including power, income, prestige, security, convenience, loyality to an idea, an institution, or the nation, pride in excellent work, desire to serve the public interest, and commitment to a specific programme of action. Thirdly, that a bureau's social function strongly influences its internal structure and behaviour, and vice versa.[13]

Anthony Downs compares the bureaus with firms. Unlike firms, however, there is no direct relationship between the service a bureau provides and the income it receives for providing them. The bureaus do not sell their goods and services in markets; the lack of output markets and the impossibility of evaluating the bureau's outputs in relation to the cost of inputs affects the bureau's operation. It receives, instead, an allocation of resources from the central budgeting agency. If the bureau is part of the government, that government collects taxes from citizens who may or may not benefit, or may even be adversely affected by the bureau's activities. There is no mechanism for matching the taxes paid by each citizen with the utility it receives from government activity, and there is a complete separation

---

[8] See Ostrom, Schroeder and Wynne 1993, pp.80-81
[9] See Weber 1947; Rowley and Elgin 1985
[10] See Tullock 1965; Downs 1967
[11] Downs 1967, p.2
[12] Ibid., p.87
[13] Ibid., p.2; 84

between the bureau's income from its supplied services. Thus, the bureau's ability to obtain income in a market cannot be used to determine how to use the resources it controls, or whether the services should be extended or maintained, nor in appraising the performance of individual bureaucrats.[14] (see Box 1.1.) The competition amongst political parties, and effective social institutions, would guarantee that the outcome produced by a bureau would represent the citizenry's preferences and would be closely related to a government's demand. Downs' model includes two crucial components, the internal and external environments of a financially dependent bureau, which both determine the outcome. The external environment, which differs with societies, predicts the power setting of a bureau consisting of those who have legal authority over the bureau (sovereign) whose power itself is limited by laws or custom. Bureaus have to compete with other organizations (rivals); they supply services to direct beneficiaries and non-beneficiaries who take an active interest in the affairs. The internal environment of bureaucratic hierarchies generates information asymmetries and distortions, and, with decisions being carried out by lower levels, interpretational scopes and discretionary powers are being used. Conflicts on one hierarchy level are resolved, if possible, on the lowest common denominator. Control opportunities for leading bureaucrats decrease with the increasing size and hierarchy levels due to rising monitoring costs.

William A. Niskanen was the first to develop a formalized model of bureaucracy's supply which departs from two important points, i.e., the lack of output markets and the self-interested bureaucrat. Niskanen developed his theory in the beginning of the 1970s when US bureaucracy was growing in size, and he was interested in the endogenous motives and dynamics of its growth. The main conclusion of his theory is that each/every publicly provided good and service tends to be expanded well beyond any tolerable level of efficiency as defined by the demands of the citizenry, and that bureaucracy plays the central role in influencing the allocation of resources.[15]

Niskanen established his theory of bureaucratic supply, analogous to the 'theory of a firm', within a monopoly/bilateral monopoly framework. He regards the relationship between a representative government (i.e., the sponsor,) and a bureau as a bilateral monopoly, where the government is the sole demander/buyer of public goods and services which are to be produced by a bureau as the sole producer/supplier. Niskanen takes it as given that the demand function for bureaucracy's services and goods is determined through the political process and political institutions, and that the government knows the preferences of the citizenry. He concedes, however, that the citizenry's demand on the bureau for services is, in any case, never directly expressed.[16] His key element is the self-interested budget-maximizing bureaucrat, who has a superior position in the budget-bargaining process over the government. Niskanen predicts 'budget maximization' as being the

---

[14] Ibid., p.30
[15] Niskanen 1971
[16] Niskanen 1973

bureaucrat's goal, because with budget maximization the bureaucrat's individual utility function can be best served (see Box 1.2.).[17]

Box 1.1. Downs' model in brief

Voters' preferences are expressed through the political process; the competition amongst political parties, and the self-interest of politicans in being 'reelected' guarantees that preferences as interpreted by a government are transferred into a government's demand on bureaucracy's supply.

A government has the dominant position in the bargaining process; self-interested bureaucrats are restricted by efficient social institutions. A government decides on the budget allocations to competing bureaus.

Characteristics of bureaucratic supply are
- monopoly supplier
- dependence on central allocations
- no direct relation between income and services
- no direct relationships between bureau and consumers.

Outcomes are determined by a government's demand.

Box 1.2. Niskanen's model in brief

Government is the sole demander/buyer on bureaucracy's supply.

Bureaucrats as budget maximizers play an active role in determining the outcome and have a dominant position over the government (monopoly, information, agenda power) in the bargaining process for the budget.

External control mechanisms are weak, because
- the bureau is a monopolistic supplier;
- there is no direct relationship between bureau and consumer;
- there is no direct relation between income and services.

Outcomes are higher than those demanded by the government and the citizenry (allocation inefficiency); the constraint for bureaucratic supply is the allocated budget from which a bundle of services is cost-effectively produced.

Bureaucrats cannot appropriate profits, because in democracies they receive fixed salaries, but they can serve their self-interested goals. The variables of the bureaucrat's utility function - income, office perquisites, power, patronage, ease of managing the bureau, ease of making changes - are, except for the last two, a positive monotonic function of the total budget of the bureau. Bureaucracies would, there-

---

[17] The 'utility function' of the bureaucrat is determined by different factors which, according to Niskanen, correlate closely to the extent of the financial budget. Budget-maximizing is supported by all members of a bureaucratic organization because career, promotion, and income expectations are positively connected to the expansion of their organization. (see Rosen et al. 1992, p.197)

fore, tend to oversupply; the outcome is somewhat higher than the government's demand.[18] While producing too much, fewer resources are left for producing other goods (allocation inefficiency), and the outcome is more costly than it would be if the goods were to be provided through private firms (X-inefficiency).

A bureau, as a monopolistic supplier of goods and services, bargains with a government for its budget. A bureau is regarded as superior in this bargaining process insofar as it possesses informational advantages over the government, and the bureau chiefs can threaten the government with zero production should it not accept the intitially proposed budget, i.e., the bureau chief makes a 'take-it-or-leave-it' budget proposal. The outcome of the bargaining is that scarce resources are inefficiently allocated, and that bureaucracies as monopolistic suppliers of goods tap the general treasury. External control mechanisms are weak because the administrators are not accountable to a bottom line demarcating benefits and costs, like those in the private sector, which leads to strongly negative net 'benefits' from many of their activities. Thus decision-makers have little incentive to consider the full social costs of their actions when this authority for decision-making is not bound together with the responsibility for outcomes.[19]

2.
Criticisms to Niskanen's Model of Bureaucratic Supply from an Institutional Perspective

Many scholars have agreed with Niskanen's central assumption on bureaucrats as budget maximizers while his model and the predicted results are, on the other hand, questioned. Critics of Niskanen focus centrally on his predicted bargaining range between the government and bureaucracy, and how the demand function for bureau services is determined, how the government decides how much it is willing to spend on bureau services, and how the government decides on the allocation of funds among competing bureaus. They criticize that Niskanen fails to devote sufficient attention to the sponsor's motivation and decision-making. He also assumes that bureaus are able to appropriate the full consumer surplus of their sponsor, but does not offer any convincing reason as to why a bilateral bargaining process should have such a one-sided outcome. The bureau's position depends on the political mechanism through which demand is expressed, and the bargaining range between sponsor and bureau cannot be properly defined.[20] In Niskanen's model,

---

[18] This is contradictory to Downs' assumption. He objected to critics who claimed an excessive growth of bureaucracy resulting from inherent tendencies not from any true social needs for service. Bureaus cannot expand without additional resources, which they must obtain through voluntary contributions from a government allocation agency. Downs assumed that the recent expansion of bureaus in democracies has occurred largely in accordance with the desires of major non-bureaucratic institutions therein. (Downs 1967, pp.257-258)

[19] See Niskanen 1981, v

[20] Hettich 1975, p.23; see also Pommerehne and Frey (1977), who introduce two other aspects: (1) they do not simply assume that bureaucracy is a monopolist, but the

the power of a bureau(crat) to obtain greater budgets than desired by the government stems from three important characteristics of the bargaining situation: „First,
the bureau is the sole supplier/producer (monopoly), second, the bureau alone
knows the cost schedule, and third, the bureau is institutionally allowed to make a
'take-it-or-leave-it' budget proposal" and can end the bargaining process with the
threat to the government of a zero production.[21] Niskanen, later, altered his own
model by introducing different variables.[22] The bureaucracy's efficiency can be
improved if - the first relaxation of the model - two or more bureaus are competing.[23] The competition between bureaus serves the sponsor with additional information about the cost schedule, and the sponsor can choose the services and goods
to be offered by the cheapest bidder. This weakens the dominant bargaining position of a bureau insofar as the government gains access to information, which is
the important second characteristic of the bureau's dominant position over the
government. The information monopoly of a bureau, furthermore, can be questioned if „the output and costs of a bureau are compared with those of a private
firm with the same type of product, or, if a bureau is the sole producer of a set of
products (...) an internal comparison at different points of time or, possibly, with
bureaus producing a simliar product in another political jurisdiction could be
made".[24] Bureaucratic competition for territory and budgetary allocations are introduced by Breton and Wintrobe as the dominant feature between specific bureaus; they recognize the possibility of residual discretionary power, which may
give rise to inefficiencies in supply.[25] Niskanen remains pessimistic about the effects of external control, because bureaus exchange their output for a total budget.
The recent literature follows his scepticism.[26] Bromley and McLean, for example,
question whether 'inefficient allocation' can be checked by external control. They
ask how a review committee can evaluate to which extent a good should be supplied, and whether it should be allocated through bureaucracy or markets. What
seems to be controllable are X-inefficiencies of different institutional arrangements.[27] The third assumption about the bureau's dominant role sets authority rules
in the bargaining situation which are in favour of the bureau. The bureau might
threaten the government with zero production ('take-it-or-leave-it' proposal), and
the authority rules would allow the bargaining by the bureau's one-sided decision

---

process of interaction with other political decision-makers, and (2) they consider institutional differences with various degrees of control of collective decisions by the voters, namely representative democracies with and without referendum, where the direct influence of the voters on political outcomes differs (pp.3-5).

[21] Mueller 1989, p.255

[22] Niskanen 1978

[23] Niskanen thus changed the boundary rules, i.e., it is no longer a bilateral monopoly
framework. [see also Ostrom (1986c), who dedicated this alternative variable to
McGuire (1979)]

[24] Niskanen 1968, pp.303-304

[25] See Breton and Wintrobe 1982

[26] Mueller 1989; McLean 1987; Rowley and Eglin 1985; Bromley 1982b; Peacock 1978

[27] See Ostrom (1986c), who examined outcomes of competition among potential producers of public goods (pp. 505), and Mueller (1989) who compared cost differences
between publicly and privately provided services (pp. 261-266).

to end. This assumption is weakened if Anthony Downs' proposition of the 'Central Role of Last Year's Budget' is introduced, or as Romer and Rosenthal argue „that the budget would revert to the status quo budget, the one used for the previous year, if the officials did not agree to the initial budgetary request".[28] Downs regards the return to the status quo as possible, because it is a reached consensus which reflects and condenses a reached power balance; it tends to create a strong degree of inertia, because many of the parties concerned wish to avoid incurring the high costs of renegotiating another agreement. Change is costly, and it seems rational to start evaluating any proposed changes by first comparing them with the status quo.[29] What is further questioned in Niskanen's hypothesis of bureaucracy's dominant position is that the bureau is institutionally allowed to make a 'take-it-or-leave-it' proposition, and itself decides when to end the bargaining process. Niskanen presumed that the bureau would have the power to decide upon the agenda, and that the bureau chief is authorized to control it. Niskanen sets authority rules which give the bureau chief full control over the agenda, and thus, over the government.[30] Peacock, on the other hand, highlights to whether the process is one of bargaining, where it is uncertain whether an agreement will be reached, or can better be described as cooperation where a consensual agreement is the envisaged result.[31]

In Niskanen's theory, the bureaucrats' utility function is crucial, and predicts, together with the assumption of the bureau's superior information position in the sponsor-bureau relationship, inefficient allocation. The Niskanen bureaucrat is a notorious rent-seeker, because the productivity of any particular investment is secondary to his private gains.[32] External control mechanisms have to be erected to limit rent-seeking, and probably corruption, and to reach efficiency. If the general assumption of self-interested rational bureaucrats is linked with opportunistic behaviour, e.g., 'self-interest seeking with guile', then „the public sector as conceptualized by public-choice theorists would not merely be inefficient, but would almost inevitably turn into an exploitative and oppressive nightmare".[33] „Efficiency", as Scharpf points out, „requires the solution of both a motivational and of an informational problem (...). Hierarchical authority is normatively acceptable only to the extent that it is exercised in the public interest, or for the common good, rather than for the benefit of self-interested power holders".[34] Scharpf follows Downs in recognizing the bureaucrats' desire to serve the public interest,[35]

---

[28] Romer and Rosenthal (1978); see Ostrom 1986c, p.500
[29] Downs 1967, p.250
[30] When Ostrom's method of institutional analysis is applied, the different outcomes of Downs' and Niskanen's model can be explained as they set differing authority and boundary rules. (see Ostrom 1986c, pp.502-503)
[31] See Peacock 1978
[32] Rent-seeking behaviour refers to decisions which can yield unearned advantages to particular individuals. (see Ostrom 1992, p.33; Tollison 1982, pp.575-602)
[33] Scharpf 1993, p.132
[34] Ibid., p.131
[35] „'The public interest' is here defined as what each official believes the bureau ought to do to best carry out its social function. Thus we are not positing the existence of any single objective version of the public interest, but only many diverse personal opinions

and he replaces the assumption of „universal opportunism with the working hypothesis according to which human behaviour is characteristically, but imperfectly, norm-oriented".[36] In the public sector, heads of government, ministers and civil servants are bound by oath of office to pursue the public interest, and he concludes that to assume that norms matter is not to deny the existence and importance of self-interested individuals, or opportunism. Scharpf argues that „it is one thing to design institutions that will reinforce the pre-existing normative orientations of agents whose own commitment to public purposes needs to be shielded against ever-present temptations, and it is quite something else to erect insurmountable walls against the ubiquitious abuse of power by agents who are assumed to be exclusively self-interested and determinedly opportunistic in their motivations".[37] As norms and norm-orientation differ among societies, different consequences could be drawn for the choice of institutions, and for the rules-in-use to set limits on opportunistic behaviour, be it corruption, or self-interest seeking with guile.

Criticisms of Niskanen have considered dependent bureaus which receive their budget from political organizations, and they have introduced variables which would increase efficiency,[38] and produce outcomes as close as possible to those desired. All these variables are based on considerations on how to improve governmental control on dependent bureaus, because it is assumed that governments 'know' the preferences of the citizenry. If 'political failures' are the potential causes for inefficiencies, and if governments fail to express the consumers' demands, then alternative models to dependent bureaus become necessary. However, Niskanen's model was to answer a specific question, i.e., the endogenous dynamics of the bureaucracy's growth which leads to inefficient budget allocations. His model can seek validity only for very restrictive conditions, i.e., a bilateral monopoly framework where one bureau bargains with a government, and where specific conditions give a bureau a superior position over a government. The object of this study, and the variables of the case study, show characteristics, e.g., the multiplicity of public, bureaucratic, civil and economic decision-makers, which are not adequately addressed by his model: in the decision-making process for public sector investments, for annual O&M budgets, for the assessment and collecting procedure of O&M water charges, and other related issues, there are more than one, and changing, political decision-makers with conflicting targets. In the case of water supply, state bureaucracies operate under uncertainty due to the nature of the resource, and information asymmetries within a bureau, and between a bureau and

---

concerning it" (Downs 1967, p.84). And „if society has created the proper institutional arrangements, their private motives will leave them to act in what they believe to be the public interest, even though these motives, like everyone else's, are partly rooted in their own self-interest. Therefore, whether or not the public interest will in fact be served depends upon how efficiently social institutions are assigned to achieve the purpose. Society cannot insure that it will be served merely by assigning someone to serve it."(Ibid., p.87)

[36] Scharpf 1992, p.133

[37] Ibid., p.134

[38] E.g., competition among bureaus supplying the same goods/services, contracting out the production to the private sector, strengthening governmental control on dependent bureaus, creating a monitoring and incentive structure for public officials.

the users, generate control and enforcement problems and give range for opportunistic behaviour. The beneficiaries exercise control only through political mechanisms (i.e., voting), and the state and its officials may respond to interventions of their lobby groups, which gives scope for distortions and inefficiencies. The state bureaucracy which operates and maintains the irrigation system is a dependent bureau of the Niskanen type, but there is more than one dependent bureau supplying services, with the private sector being involved in the execution of investments. A local branch of a central bureaucracy supplies services, and two types of water-user organizations contribute towards operation and maintenance activities with changes in the supplier-demander structures where the beneficiaries/consumers exercise varying degrees of control.

3.
The New Institutional Economics, or the Institutional Rational Choice Approach

Unlike Niskanen and other Public Choice theorists, scholars of the New Institutional Economics[39] „seek to explain characteristics of social outcomes on the basis not only of agent preferences and optimizing behaviour, but also on the basis of institutional features".[40] The New Institutional Economics attempts to overcome deficiencies in the mainstream neoclassical economics which have considered individual preferences, technological opportunities, physical and capital endowments and market opportunities as constraints to decision-making, and where institutional variables have been taken as given. Although there is no agreement in defining 'institutions', there are some common characteristics which refer to the nature of their rules and constraints, their ability to govern the relations among individuals and groups (e.g., whether they are voluntarily accepted through custom and tradition, or are enforced and policed through an external authority and a coercive incentive system, they are, however, applicable in social relations) and their predictability (e.g., they are applicable in repeated and future situations, and provide some degree of stability).[41] One strand of the NIE scholars regards institutions as complexes of norms and rules which direct behaviour to serve a collective purpose. Institutions can be defined as rule-ordered relationships among people, or as Shepsle said, „institutions may be thought of as part of what embeds people in social situations. They are the social glue".[42] Elinor Ostrom uses rules as a referent for the term 'institution', and institutions are defined by rules which are not identical to formal laws or a set of written rules, although they might be. She regards rules as „potentially linguistic entities that refer to prescriptions commonly known and used by a set of participants to order repetitive, interdependent relationships. Prescriptions refer to which actions are required, prohibited, or permitted. Rules

---

[39] The term 'New Institutional Economics' is used here as the general banner under which all concerns with institutional features have been elaborated.

[40] Shepsle 1989, p.135;137

[41] See Nabli and Nugent 1989, p.1335

[42] Shepsle 1989, p.134

(...) are distinct from physical and behavioral laws. Rules are the means by which we intervene to change the structure of incentives in situations. Theoretically, rules can be changed while physical and behavioral laws cannot. Rules are interesting variables precisely because they are potentially subject to change. That rules can be changed by humans is one of their key characteristics".[43] Institutions shape human behaviour through their impact on incentives, which are understood as the positive or negative changes in outcomes that individuals perceive as likely to result from particular actions taken within a set of rules-in-use, combined with the relevant individual, physical, and social variables that also impinge on outcomes. Incentives are not only financial rewards or penalties, but „opportunities for distinction, prestige, and personal power; desirable physical conditions of work; pride in workmanship, service for family or others, patriotism, or religious feeling; personal comfort and satisfaction in social relationships; conformity to habitual practices and attitudes; feeling of participation in large and important events".[44] Institutional arrangements establish the incentives, information and compulsions that guide behaviour and determine economic outcomes; they establish the basis for market control or administrative control, or the basis for contractual agreements between independent principals to govern their own affairs, based on bargaining, cooperation and persuasion. Institutions, themselves, are a choice, subject to endogenously, or exogenously, induced transformations.

The New Institutional Economics is not a homogeneous concept, or theory, and there is no consensus on what is included in this approach; but the axiom to which they refer is that of self-interested individuals whose rationality is bounded, and whose behaviour includes opportunism. The label New Institutional Economists refers to theorists who assume that transaction and information costs strongly influence the efficiency of institutions, and the selection of institutions. NIE scholars make the point that information and transaction costs are important in explaining the likelihood of cooperation, the competitiveness of alternative forms of organizations, the effectiveness of collective action, and the role of a state.[45] Recognizing the relevance of transaction and information costs, and the problem of collective action and free-riding behaviour are, however, two central issues of the NIE approach.[46]

3.1
Transaction and Information Costs

A common hypothesis of the approaches of the New Institutional Economists is that institutions are transaction cost-minimizing arrangements. They may change and evolve with changes in the nature and sources of transaction costs and the means for minimizing them.[47] Transaction costs reflect the fact that decision-making and performing activities are not without costs. Transaction costs are as-

---

[43] Ostrom 1986b, p.5
[44] Ostrom (1992) quoting from Simon, Smithburg and Thompson (1958, p.62)
[45] De Janvry, Sadoulet and Thorbecke 1993, p.568
[46] See Nabli and Nugent 1989, pp.1336-1340
[47] Ibid. 1989, p.1336

sociated with the search for information, the process of bargaining, and the legal enforcement of contractual agreements.[48] Transaction costs are prominent in the search for contract partners, as well as the preparation and conclusion of contracts on the exchange of exclusive, transferable rights of goods. They, in general, evolve in situations where individuals act interdependently, be it with, or without, institutional restrictions. Incomplete information and asymmetries in information can be considered as one source of transaction costs, and the search for information and monitoring is necessary to minimize losses for a party to a contract. The conditions for the market to fulfill its function, despite the occurence of transaction costs, are well-defined property rights which institutionally determine the range of discretion of the economic agents. The connection between property rights and transaction costs is picked out by Alchian and Demsetz as a central theme because transaction costs are regarded as having important impacts on the allocation of resources.[49] Technological, together with other conditions, including externalities, can give rise to the kinds of institutional mechanisms for internalizing externalities, known as property rights, which Alchian and Demsetz define as a 'bundle of rights', e.g., what one possesses are socially acknowledged rights to use.[50] The existence of property rights may reduce conflicts and facilitate cooperation, resulting in a reduction of transaction costs. In this way, along with technology and other constraints, institutional constraints enter into the decision process of individuals. Transaction costs also emerge in the public sector if one considers that political and administrative decision-making and the implementation of decisions are distinct from each other. Governments and bureaucracy are plagued with internal sources of inefficiency arising from transaction and information costs. Transaction costs can reach prohibitive dimensions where they impede interaction; alternative coordination mechanisms, or institutions, are then desired.

A general 'agency problem' in contractual agreements is recognized in both the economics of transaction costs and that of costly information, which is systematically reflected in the Principal-Agent approach.[51] The Principal-Agent approach deals with the separation of ownership and management,[52] and is applied to situations where contracts are incomplete and/or information is asymmetric. All relationships in which one individual depends on the action of another, e.g., the individual taking the action is called the agent, the affected party the principal, are characterized by transaction and information costs.[53] Due to asymmetric information in uncertain environments, incentive schemes and control mechanisms become necessary. The agency theory now makes the point that the loss which the principal suffers, in comparison to the situation of free, complete information, due to the

---

[48] See Weigel 1987, p.61; 65-68
[49] See Alchian and Demsetz 1973
[50] Alchian and Demsetz 1973, p.175
[51] Nabli and Nugent 1989, p.1337
[52] See Rowley and Elgin 1985; Pratt and Zeckhauser (eds.) 1985; Tirole 1988; Holmstrom and Tirole 1989
The Principal-Agent approach challenged the view of Demsetz and Alchian with the recognition that the classical capitalist firm changed with corporate enterprises and with the separation of ownership and control.
[53] Pratt and Zeckhauser 1985, p.2

existence of asymmetrical information and the uncertainty of the results of actions, is as low as possible. The agency costs should be minimized through the design of contracts.[54] The literature on principal-agent relationships focuses on the structure of an agreement, i.e., a contractual, rule-ordered relationship, that could induce agents to serve the principal's interest even when the agent's actions and information are not, or cannot be, observed by the principal. Because the principal cannot perfectly, and without cost, monitor the agent's action and information, the need for an arrangement arises where „the principal and agent have a common interest in defining a monitoring-and-incentive structure that would produce outcomes as close as possible to ones that would be produced if information and monitoring were costless".[55] Many instruments are available to reward agents, not only pecuniary, but „social rewards, such as reputation in the community or the returns that come from adhering to a set of ethics. Friendship, family, and connection also play a substantial role in creating additional types of incentives".[56] The Principal-Agent approach also applies to the relationships between the government (principal) and bureaucracy (agent); like the preferences of the shareholders (the principal) and the managers (the agent), those of the government and the bureaucracy do not coincide. The government is faced with trying to control the unobservable actions of a bureau. Alternative institutions can substantially reduce monitoring and control costs, i.e. transaction costs. Decentralization can change the setting by creating direct accountability linkages between the beneficiaries (principals) and the bureau (agent). If the beneficiaries' payments for goods and services constituted the budget of a bureau, then their willingness to pay for the goods and services could be expressed, and could become an incentive to improve the performance, i.e., supplying services according to the beneficiaries' preferences; but even then, monitoring and control is costly, and collective action problems arise if there are numerous interdependent principals.

3.2
Collective Action

The key issue in the collective action literature, i.e., the second central theme of the NIE, is to explain collective outcomes in terms of individual motivation, or „to explain the likelihood of success or failure of a given set of self-interested individuals in undertaking actions that may benefit them collectively".[57] This concerns public or collective goods, and, particularly, the use of common-pool resources, where the exclusion of self-interested individuals who do not participate in the provision is difficult, or even impossible, and where rivalry in consumption produces externalities. The incentive for individuals to free-ride is analyzed by Mancur Olson and Garrett Hardin, who assumed inefficient collective outcomes in that public, or collective, goods are underprovided, and/or overused unless there is a system of individual property rights. This pessimism on cooperation, which is

---

[54] Schumann 1992, p.456
[55] Pratt and Zeckhauser 1985, p.6
[56] Arrow, quotet in Pratt and Zeckhauser 1985, p.17
[57] Nabli and Nugent 1989, p.1338

theoretically reflected in the prisoners' dilemma game, is based on the assumption that non-cooperation is the dominant strategy of individuals. However, as Taylor and others have pointed out, the constellation of costs and benefits of collective action is much more favourable to the possibility of cooperation than the prisoners' dilemma game.[58] In the case of common-pool resources, such as irrigation systems, collective action may benefit individuals more than free-riding. A growing body of literature is concerned with the conditions and determinants for successful cooperation,[59] and recent studies have shown that common-property institutions can overcome social dilemmas, and can achieve efficiency in the use of natural resource systems by voluntary cooperation.[60]

The theory of rent-seeking which is receiving increasing attention in the collective action literature, addresses the relationship between the state and interest groups. It is presumed that the state and its officials are not neutral in the process of group interaction, and that they do not only passively facilitate, or mediate, collective action.[61] If public officials and lobbies benefit from the predatory behaviour of interest groups, then institutional change could be introduced to limit lobbying and to shield civil society from its own actions.

4.
The Nature of Water as a Resource, and the Peculiarities of Irrigation and Drainage Systems

Government intervention is regarded as necessary if exclusion of potential beneficiaries is unfeasible and if consumption is non-rival, as it is with pure public goods which have two salient characteristics, the jointness of supply, and the impossibility of excluding others efficiently from its consumption. A pure public good or service is defined as being available to all members of the appropriate community. A large number and variety of consumption units, all of which are somehow identical, are provided by a single one of the goods, as it is produced. It will not, however, be competent in excluding any person from the positive or negative enjoyment of its availability once it is produced. Here, according to Musgrave, the principle of the exclusion characteristics of market-produced goods breaks down. In the extreme or directly opposing sense, non-exclusion applies. Additional consumers may be added at zero marginal costs.[62] National defence, law and order, and clean air are the classic examples for pure public goods. Unlike private goods, when exclusion is feasible and preferences are expressed by the willingness to pay for this or that good (quid pro quo transactions), public goods and services yield non-excludable benefits. Failure of the exclusion principle to apply provides an

---

[58] From Bardhan 1993, p.634
[59] For the determinants of sustainable cooperation, see Elinor Ostrom's set of rules (in Chapter Two, 2.); de Janvry et al. 1993, pp.568-569
[60] See the following Chapter Two, 2. Small-scale farmer-owned and -managed irrigation systems
[61] See Repetto 1986
[62] Buchanan 1968, p.49.

incentive for non-cooperative, individualistic behaviour, or free-riding.[63] Non-exclusivity is not the cause for free-riding but the absence of an institution for exclusion that controls the pursuit of individual behaviour. The number of benefiting individuals may differ extensively, but even in small groups the incentive to free-ride is present at the expense of others, e.g., the non-contribution to the provision of the goods, because it is unlikely that their contribution will bring them benefits that exceed their costs. Or, as Mancur Olson expressed it, if the number of those contributing increases, the return to the individual on its additional contribution becomes smaller, and the incentive to free-ride increases.[64]

Irrigation and drainage systems share characteristics with public and private goods, and these characteristics have important implications for how they should be provided and maintained, and for their operation.[65] When surface water resources are made available for agricultural purposes through the construction of irrigation and drainage systems, they introduce two types of physical linkages among all, formerly independent, farmers receiving their water from one supply system within a drainage area. The interdependencies refer to the joint use of a water delivery system from which farmers appropriate resource units as factors of production, and to the joint use of a drainage system. The first interdependency generates provision and maintenance problems for the delivery system, and problems on the allocation of water resource units where interfarmer conflicts arise. The use of water as an input factor in farm economy creates the necessity for the provision and maintenance of drainage systems to discharge the effluents from/of private farm-firms, e.g., the second interdependency, and for other groundwater and salinity control means (see Fig. 1.1., Interdependencies in a surface irrigation system).

---

[63] „It is worth noting that there are no *personal elements* in the individual's calculus of decision here, and, for this reason, the 'free rider' terminology so often used in public-goods theory is itself somewhat misleading. The individual is caught in a dilemma by the nature of his situation; he has no sensation of securing benefits at the expense of others in any personal manner. And to the extent that all persons act similarly, no one does secure such benefits. The Prisoners' Dilemma is more descriptive of the large-number behavioral setting. In such models, non-optimality arises because of the mutual distrust and non-communication between the prisoners. In the large number or n-person dilemma, the failure to attain desirable results through independent action is analytically equivalent to the orthodox Prisoners' Dilemma. The organization and enforcement of efficient institutional arrangements will rarely be possible unless all persons are somehow brought into potential agreement. The alternative of remaining outside the agreement, or remaining a free rider, must be effectively eliminated." (Buchanan 1968, pp.86-97; in the original without italics)

[64] See Olson 1968

[65] See The World Bank 1994, p.25, Fig. 1.3 („Infrastructure services differ substantially in their economic characteristics across sectors, within sectors, and between technologies") does not consider drainage infrastructures.

4.1
The First Interdependency: the Joint Use of the Delivery Infrastructure and the Use of the Resource

A surface irrigation system can be considered as a common-pool resource which shares characteristics with private and public goods.[66] All eligible beneficiaries use the technical facility in common (joint-use, non-separability), but the resource units which individuals appropriate are subject to private use. The available water is limited, and each individual's use substracts resource units from those available to others: if the upstream farmer A substracts water, the water availability for the downstream farmer B changes. Individual farm-firms lose control over factors of production, e.g., technological or physical externality,[67] and thus different production functions exist for farmers along a watercourse. One party is able to acquire a form of property right over the income stream of others which water can create. This interdependency is unlike that of the normal market (pecuniary externality). If the impact on the production function of another individual firm is not fully accounted for through the price system or other transactions, a Pareto-non-optimal allocation of resources results. Or as Bromley explains, „in irrigated agriculture we have a situation in which downstream irrigators receive water at the discretion of those upstream on the watercourse (and) independent economic activity leads to both inefficiency and inequity".[68]

The exclusion of potential beneficiaries from the access to the resource is difficult, either because of the size of a delivery scheme, or because the costs of exclusion are higher than the potential benefits, or simply because exclusion is socially not desired. The provision of the technical facility and its sustainment, i.e., its maintenance, generates problems similar to the provision of pure public goods, which suffers from free-riding, because it is difficult to monitor and to prevent access.[69]

The first characteristic, i.e., the separability of resource units and the costs of exclusion, results in conflicts over the consumption of water if no accepted rules for water allocation, and no accepted systems of control are in place. Due to the substractable nature of the resource flow, rule-ordered allocation procedures are necessary to make optimal use of the available resource. Water allocation rules should ensure adequate and reliable supply in quantity, quality, location and timing to respond to the individual farmers' demand. The availability of water is an uncertain matter, and although the storage of surface water in reservoirs enhances security of water availability, water supply is still determined by the nature of the hydrological cycle, showing high variabilities within a year and from year to year.

---

[66] Ostrom 1990, pp.30-33; Ostrom et al. 1994; Tang, in: Ostrom et al. 1994, pp.225-245
[67] Physical externalities have been recognized by economists as a potential source of market failure which impedes efficient resource allocation. The terms 'externality' or 'non-exclusivity' label the same problem.
[68] Bromley 1982a, 1.05
[69] See Ostrom 1992

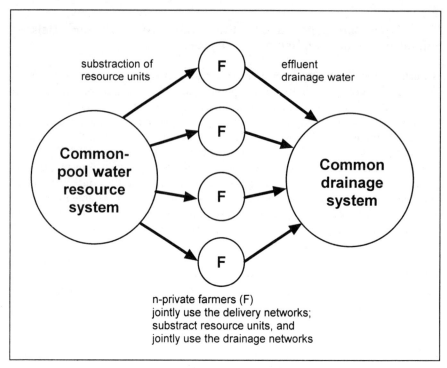

Substraction of resource units — effluent drainage water

Common-pool water resource system

F
F
F
F

Common drainage system

n-private farmers (F) jointly use the delivery networks; substract resource units, and jointly use the drainage networks

Fig. 1.1. Interdependencies in a surface irrigation system

Therefore, guarantees of a particular amount of water at all times cannot be given, but the probability of water availability and the allocation procedures under changing circumstances should be known to the users,[70] though the farmers may react to environmental uncertainty in an institutionally certain environment. Water allocation rules are like a constitution within an irrigation system; they define (a) who has a legal claim to use the resource and the resource system, and the obligations therein, (b) varying procedures regulating the appropriation of resource units among a large number of users in times of water abundance and scarcity, (c) whether the rights are transferable property or usufruct rights, (d) who has to bear what costs, and (e) sanctions on rule opposers. Without established and enforceable water allocation rules, the right to use water would be gained through its capture from channels following the principle 'first come, first served' to the detriment of other users. This situation would resemble the Prisoners' Dilemma where the individuals' behaviour is characterized by non-cooperative strategies with negative payoffs, and open access to the resource would be significant. Unilateral behaviour of the irrigators in a situation of physical interdependence produces inefficient outcomes. The second characteristic, i.e., the non-separability and

---

[70] See Livingston 1992

costly exclusion from the use of the infrastructure system, creates the need for enforceable provision and maintenance rules to limit free-riding.

4.2
The Second Interdependency: the Joint Use of the Drainage Infrastructure, and Externalities Caused by Agricultural Effluents

The physical characteristics of irrigated agriculture produce external effects through rising groundwater tables and the associated salinization of soils imposing costs on other users.[71] It therefore requires groundwater level and salinity control means, because even if water is used efficiently, salt has to be leached, and excess water needs to be discharged out of the area. Although decreased water inputs into the area can reduce drainage requirements, drainage systems are the most important technical prerequisite to sustain agricultural production over the long term. A man-made drainage system is a good which shares no attributes with private goods but all with public, e.g., non-separability and non-excludability. Drainage has public good characteristics but they are not strictly purely public, or collective, goods because „the amount of benefits that an individual receives is readily identifiable, and is largely in proportion to their landholding".[72] Small and Carruthers consider drainage to deserve the status of a merit good. Merit goods are provided by a government above the level indicated by the market because the public may be unaware of the real benefits of an investment. Then government involvement may be justified, as it is with drainage. The lack of it would show that merit wants need to be given serious considerations.[73] Due to the nature of the drainage systems, irrigators in a project area cannot be excluded from their use, or at least at high costs. The public nature of the good 'drainage system' generates provision and maintenance problems, and the incentive to free-ride, i.e., the non-contribution to provision and maintenance, is more striking than for the provision and maintenance of delivery systems. This originates in differently perceived costs and benefits, because the negative consequences of non-provision and non-maintenance can emerge with substantial time lags, and are spatially inhomogeneously distributed.

In common-pool resource systems, the combination of the two factors, i.e., non-excludability and substractability, can result in situations which Garrett Hardin described as 'the tragedy of the commons'. Howevver, common dilemma situations do not necessarily result in the deterioriation of common goods if both types of interdependencies among economic farm units are mediated through enforceable rules for water allocation, and for the provision and maintenance of irrigation and drainage infrastructure facilities.

---

[71] This is, of course, only true for cases which are relevant to this study.
[72] Small and Carruthers 1991, p.28
[73] See Small and Carruthers 1991, pp.28-29

4.3
The Mediation of Externalities in Common-Pool Water Resource Systems Under Different Property-Rights Regimes: State - Market - Civil Organizations

Scholars of the property-rights approach make a distinction between the natural resource itself and the property-rights regime under which it is held.[74] They assume that „there is nothing inherent in a resource itself to determine <u>absolutely</u> the nature of property rights. The nature of property and the specification of rights to resources are determined by the members of a society and the rules and conventions they choose to establish - not by the resource itself".[75] Surface irrigation water, and the technical facilities, may be held under more than one property-rights regime, and Bromley namely labelled state property regimes (res publica) where neither the managers nor the users are the owners, private or individual property, and common property regimes, [i.e., private property for a group (res communes)], which form a third class; they all are distinct from open-access situations (res nullius) where no property rights exist (see Fig. 1.2.).

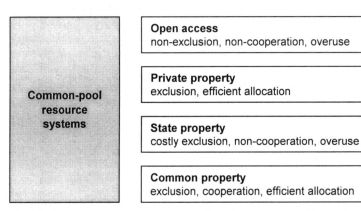

Fig. 1.2. Common-pool resource systems under different property-rights regimes

Government initiative in both investment and supplying services is a means by which collective action can be initiated, and a problem of major concern is if the state can ascertain collective action among interdependent water users to mediate the externalities. The most common approach in large-scale irrigation projects to water allocation is centralized governmental control embodying the externalities so that a single management unit will fully internalize the external effects, and the public provision and maintenance of the delivery systems. However, as experiences in many large-scale irrigation schemes suggest, they resemble more open access situations. The mediation of both types of interdependencies by governmen-

---

[74] See Ostrom 1990; Bromley 1989; Gibbs and Bromley 1989; Berkes 1989
[75] Gibbs and Bromley 1989, pp.24-25

tal actions is criticized for producing most, if not all, of the failures which under private property rights would not have occurred.[76]

With state intervention, there would be a systematic understatement of the prices of publicly provided and produced goods and services, and bureaucrats, free from the discipline of the market, could ignore prices that the market would otherwise have placed. Baden and Stroup, for example, analyzing the effects of bureaucratic management on environment, favour private property rights as the better tool of resource management, because an imperfect market might actually be the best available alternative, creating proper incentives for economic efficiency.[77] If private property rights are clearly established and enforceable, individual decision-makers would accurately reflect the social costs of their decisions, and external effects could become internalized. This solution to externalities derives from Coase, who assumed that market-oriented solutions to externalities are possible if private property rights to common goods are clearly defined and transaction costs are zero.[78] Property rights are neutral when there are no transaction costs, i.e., are without influence on allocation; all negative effects are internalized, because during the direct negotiations between the involved participants the classification of goods will be a successful one which, after payment of compensation and default costs, presents the highest possible added value. All exchange possibilities which promise mutual advantage are realized. An efficient system of property rights comprises universality (that is „all resources should be owned or ownable, by someone, except resources so plentiful that everybody can consume as much of them as he wants without reducing consumption by anyone else"), exclusivity which eliminates free-riding, and transferability, because if property rights cannot be transferred, there is no way of shifting a resource from a less productive to a more productive use through voluntary exchange.[79] These conditions, however, are very restrictive, and the physical nature of water poses problems on water markets: it is difficult to specify units of water, and measurement problems occur. According to Small and Carruthers, „another constraint is the fact that irrigation water is seldom fully 'used' by farmers. Most water users only consume a part of the available supply and the rest becomes accessible to downstream users".[80] Key requirements for water pricing and, thus, water markets, are, however, rarely met, e.g., selective water deliveries and measurement which does not necessarily mean volumetric.[81] Water use and allocation involve, however, group decisions and actions, and, thus, water transfer through markets has to overcome substantial coordination costs imposed by the informational problems among irrigators, and substantial enforcement, i.e., transaction costs.[82] Difficulties arise when the costs imposed on other irrigators have to be accounted, because yield losses cannot be specified on an accurate data basis. It cannot be determined who will have to pay

---

[76] See de Janvry, Sadoulet and Thorbecke 1993
[77] Baden and Stroup 1981, p.8; see Baden and Stroup 1983; R. Hardin 1982
[78] Coase 1960
[79] Posner 1977, pp.10-13
[80] Small and Carruthers 1991, p.36
[81] Ibid., pp.85-89; 95
[82] See Livingston 1992, pp.16-17; Saliba and Bush 1987

what compensations. If the individual farm units tried to bargain with others over the costs which derive from salinization, hundreds of thousands of distinct pair-wise externalities would be involved, and agreements between two individuals might be upset by the actions of a third.

Scholars of the New Institutional Economics share the scepticism on the public provision and supply of goods and services, but they oppose the simple diagnosis 'state failure' and the need for privatization, or 'market failure' and the need for governmental control as it is expressed by Elinor Ostrom in Governing the Commons".[83] She studied traditional self-governing institutions managing a variety of common-pool resources where the individual's decision-situations have produced efficient outcomes. Irrigated common property regimes, i.e., farmer-owned and -managed systems, are able to overcome social dilemmas, i.e., the discrepancy between the rational behaviour of individuals and the failure to attain the social optimum.[84] This is in opposition to the conclusion of Garrett Hardin's analysis that all situations where individuals jointly use a resource inevitably lead to its deterio-riation.[85] It is assumed that these regimes provide a structured bargaining forum that limits the costs for coordination, and transaction costs can be reduced by es-tablishing rules that are less costly to enforce. Customs and conventions induce cooperative solutions that can overcome collective action difficulties and help to achieve efficiency in the use of the resource.

5.
A Concept for Analyzing High Groundwater Levels and Salinization in Large-Scale Public Irrigation Systems

The conceptual unit of Elinor Ostrom's method of institutional analysis is called an action arena, including an action situation component and an actor component (see Box 1.3.). The rules-in-use form the structure of the 'action situation', and offer explanations of actions and results. Decision-making is understood as the outcome of strategic games between multiple participants, all of whom have preferences and well-specified objectives, and are incentive-constrained by means of an incen-tive scheme and by the type of compliance and cooperation mechanisms. Out-comes appear to track preferences of distinguished actors upon whom institutional structure and procedure conferred disproportionate agenda power.

Ostrom defines an action arena as a setting in which a certain type of action takes place; this includes informal as well as formal settings such as legislature and courts. Rarely corresponding exclusively to a single set of rules, several collective-choice arenas can frequently affect the set of operational rules.

---

[83] Ostrom 1990

[84] Social dilemmas are theoretically reflected by the Prisoners' Dilemma: incomplete information leads to the non-cooperative behaviour of rationally acting individuals, and they fail to reach an optimum that can be attained by cooperation.

[85] G.Hardin (1968) considers that no 'invisible hand' directs the events for the best of everyone. By following their own interests, individuals of a society which believes in the free use of the commons move towards the ruin of everybody.

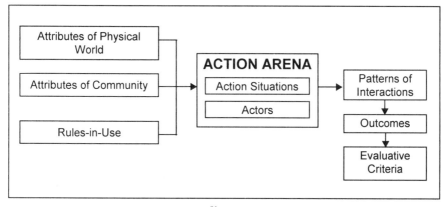

Fig. 1.3. Framework for institutional analysis[86]

Box 1.3. Components of action arenas[87]

An Action Situation involving
Participants in
Positions who must decide among diverse
Actions in light of the
Information they possess about how actions are
Linked to potential
Outcomes and the
Costs and benefits assigned to actions and outcomes.

Actors, the participants in Action Situations who have
Preferences,
Information-processing capabilities,
Selection criteria, and
Resources.

Informal and formal constitutional-choice proceedings may likewise take place in local, regional, national and/or international arenas.[88] Action situations refer to the social space where individuals interact, and actors are the participants in a situation whose behaviour is to be explained. An action situation is, for example, what has been called a Commons Dilemma situation, and Ostrom states that „once an action situation is identified as having the structure of a commons dilemma situation, predictions about likely behaviour can be made no matter whether the participants are relating to an ocean fishery, an overcrowded bridge, a meadow owned in common (...). Unless participants have worked out an implicit or explicit contract for allocating the use of the commons, an analyst would predict that each will be

---

[86] Source: Ostrom 1992, p.37
[87] Source: Ostrom 1992, p.29
[88] Ostrom 1992, pp.46-47

led to adopt a strategy that leads all of them to overuse the commons".[89] Factors affecting action arenas are attributes of the physical world, attributes of the community, and the rules-in-use.[90] The attributes of a community are of particular importance as they affect the structure of an action arena including „the norms of behaviour generally accepted in the community, the level of common understanding that potential participants share about the structure of particular types of action arenas, the extent of homogeneity in the preferences of those living in a community, and the distributions of resources among those affected. The term 'culture' is frequently applied to this bundle of variables".[91] She describes seven classes of rules by which an action arena is configured (see Box 1.4.). These rules are variables which are common for diverse types of institutional arrangements. A change in one of these variables produces a different situation, for example, by introducing more participants, or changing the set of alternative actions.[92]

Box 1.4. Rules configuring an action arena[93]

Boundary Rules set the entry, exit, and domain conditions for individual participants.

Scope Rules specify which states of the world can be affected, and set the range within which these can be affected.

Position Rules establish positions, assign participants to positions, and define who has control over tenure in a position.

Authority Rules prescribe which positions can take which actions, and how actions are ordered, processed and terminated.

Information Rules establish information channels, state the conditions for when they are to be open or closed, create an official language, and prescribe how evidence is to be processed.

Aggregation Rules prescribe formulae for weighing individual choices and calculating collective choices at decision nodes.

Payoff Rules prescribe how benefits and costs are to be distributed to participants in positions given their actions and those of others.

When Elinor Ostrom's concept is applied to analyze institutional constraints in public irrigation systems which produce, or do not account for, negative externalities, it is useful to differentiate between a series of decision-making and action. Their final results are the provision of the good 'irrigation and drainage infrastructure' and the supply, or production, of a bundle of services such as water allocation, maintenance and salinization control.[94] The first action arena is the planning and design process. In the second arena provision decisions are made; there,

---

[89] Ostrom 1986a, p.460
[90] See Ostrom, Gardner and Walker 1994, pp.23-50
[91] Ostrom 1986, p.472
[92] See Ostrom 1986a, pp.462-464
[93] Source: Ostrom 1986a, p.468
[94] In order to determine the role of government in water resources management, provision and production are distinguished. (see The World Bank 1993, p.85)

elected public officials bargain with bureaucracies over the amount of the budget to be allocated, if goods and services should be provided and how they should be financed, the quantity and quality of the goods and services, the degree to which private activities related to these goods and services are to be regulated, how to arrange for the production, how to finance the provision, and how to monitor the performance of those who supply these goods and services.[95] Production decisions, i.e., the act of executing investment and generating O&M services,[96] are taken in the third and fourth action arena where transformation and transaction activities are relevant.[97] Analyzing dependent state agencies, it seems to be important how they are constituted, what authority they have, how limits on their authority are maintained, how they obtain information, how their agents are selected, and what incentives they provide. As the beneficiaries jointly use the irrigation and drainage infrastructure, and individually substract resource units, the inducement of collective action is important. The services are supplied to individual private farm-firms that try to maximize the benefits from water. Decisions in farm economy constitute a fifth action arena where water resource units serve as a central input factor to which others have to be adjusted to gain maximum advantage from them, and where agricultural effluents are produced. There, demand for salinization control arises, and it becomes important how it can effectively be expressed.

---

[95] ACIR (1987) quoted in Ostrom et al. 1993, p.74

[96] Or, as Ostrom et al. (1993) define, it is „the more technical process of transforming inputs into outputs - making a product, or, in many cases, rendering a service" (p.75).

[97] Transformation activities are directed towards changing one state of affairs into another; transaction activities are directed towards the coordination of transformation activities, the provision of information, and the acquisition of a strategic advantage over others.

Chapter Two
Literature Review on Large-Scale Public and Small-Scale Farmer Irrigation Systems

The poor performance of many public irrigation schemes has demonstrated the weakness of government intervention, and some analysts, particularly neo-liberal economists, have denounced the positive governmental actions in natural resource management for producing most, if not all, of the failures which under private property rights would not have occurred. Privatization of the resource would result in excludability and an internalization of the externalities, and would allow market forces to achieve more efficiency.[1] The nature of irrigation and drainage systems creates, however, numerous sources of market failures. Empirical research on traditional farmer-owned and -managed irrigation schemes have shown that these systems can guarantee an efficient water allocation and that they sustain the irrigation infrastructure. In many large-scale public irrigation systems, water-user organizations contribute towards operation and maintenance activities, and analysts have conceded that they have a positive impact on performance, but these studies have remained unsystematic as there is a wide range of options for joint management, and no studies are available as to whether they show a higher potential for controlling salinization than public systems.

High water inputs into the area and the insufficient drainoff of excess water, i.e., the two central factors generating high groundwater levels, waterlogging and salinity, are associated with investment and maintenance decisions, with operation activities, the financing of irrigation services, and with the use of water at on-farm level. While the relation between investments in delivery and drainage systems, the impact of maintenance activities and the emergence of high groundwater levels and salinity is unquestioned and welldocumented,[2] it is not immediately apparent in which fashion the operation activities contribute. Excess water inputs, or 'overirrigation', are usually attributed to the farmers' unawareness and to their lack of experience with irrigated agriculture, despite the fact that excess water applications are also reported in long-running irrigation systems where farmers have gained expertise. The common consequences are to emphasize the necessity of agricultural extension services, which are without doubt important, especially when the farmers change from dry-farming to irrigation. Economic approaches interpret high groundwater levels and salinity as the result of inefficient water use, and the high external costs associated with agricultural drainage have added new impetus to improve water-use efficiency. Water is regarded as an increasingly

---

[1] See Stroup and Baden 1983; R.Hardin 1982
[2] See 1.2 in this chapter

scarce and valuable resource, and its price should reflect its opportunity costs, and the external costs incurred. In most of the countries all over the world, however, farmers do not even pay the working costs of public irrigation systems, and area-based water charges provide no incentives for water saving. Scholars assume that efficient water allocation and efficient water use can be promoted, or discouraged, by the enforcement of water allocation rules as the central part of the operation activities.[3] As the water allocation rules and their enforcement may positively influence water-use efficiency, they may, nevertheless, be a source of high water inputs.[4]

One important point in this chapter is, however, how water allocation rules and operation activities influence efficient water allocation and water-use efficiency at on-farm level in large-scale public, in farmer-managed, and in jointly managed irrigation systems; it focuses on how the management units deal with the negative externality of high groundwater levels and salinity, and on their difficulties and potentials for supplying the service 'salinization control'.

1.
Large-Scale Public Irrigation Schemes

Empirical studies on *large-scale public* irrigation schemes are selected for review which show, foremost, a certain type of decision-making: financially dependent state agencies are the most common *type of decision-making bodies* in large-scale public irrigation schemes. They are embedded in a highly centralized institutional structure which has important implications on the provision, operation and mainte-nance, and use of water. Their constraint for supplying services is the centrally allocated budget, but it is presumed that non-budgetary constraints are equally important. Multiple participants are involved, each having different preferences and constraints, different perceptions of costs and benefits and a differently dis-counted future. Those deciding on the provision and its financing are not the deci-sion-makers who operate and maintain it, and they are not the same as those who use the facility and the natural resource. Direct linkages among those making the decisions on provision and supplying services, those benefiting from the decisions, and those bearing the costs do not exist. The question of *size*, i.e., the service area covered, cannot be defined in a simple manner although this seems to be more easy. Examples in the literature range from several hundreds of ha to some thou-sands and hundreds of thousands of ha. The major point is that with larger service areas and an increasing number of farm units, operation activities, in combination with the particular water distribution methods, become more difficult and complex with increasing information, coordination and transaction costs. Maintenance re-quirements subsequently rise, and the implementation of maintenance activities faces similar difficulties.

---

[3] Small et al. (1989) state that in South Korea and the Philippines, for example, ineffi-ciencies in water use are related to the ineffectiveness of the management more than to pricing.

[4] See Chambers (1988) and Bottrall (1981), who focus on main system management

1.1
Water Allocation Rules and Operation Activities

Water allocation rules in public irrigation systems assign user rights for water to all farmers in a project area, and their assignments are subject to landownership or leasing.[5] Water is supposed to be available for all plots within a project area, but sometimes more irrigators have user rights than the source of water can support. Usually, user rights are not transferable, and the principle of 'use it or leave it' governs.[6] In many countries, the beneficiaries only have to pay for the provided services, and in a few countries they contribute a share to the initial investment costs. Fines on non-payers are levied, but non-payments have rarely any consequences on user rights. Low sanctions, if ever applied, are foreseen on those who oppose the water allocation rules. User rights are not invariably set, and they change in times of abundancy and scarcity. In times of seasonal low water, the assignments are more restrictive, but there are no compensations for those farmers who receive no water.

The substantial, i.e., the real/de-facto, rights to use water depend on, and change with, water availability, the design of the conveyance systems, the chosen water delivery methods, and, last but not least, on the quality of services provided. The management unit of an irrigation project conducts numerous operation activities to implement the farmers' rights to water, e.g., planning, daily implementation, monitoring and control,[7] and it is the operation activities which finally determine the de-facto rights. The operation of a system renders necessary the coordination within a management unit, and the cooperation between the unit and thousands of farmers, and it depends on the efforts and capacities of the involved persons, if the water supplies are adequate, reliable and predictable. The implementation of operation activities is not only a problem of sufficient or insufficient funding, but is conditioned by a variety of factors (of course, some of them are a function of funding), e.g., skilled personnel, technology, how information is processed, and how coordination, cooperation and internal monitoring are achieved.

Bottrall points to the important implications which the water delivery methods have on supply. Each distribution method has its own characteristics and may, or may not, suit local conditions, and great efforts are required for each method to solve communication and coordination problems between the large number of farmers and the managing unit, and within the operational units themselves (see Box 2.1.).

---

[5] Water allocation rules within an irrigation system are different from a water rights system which should apply to different water sources, categories of use, quantity and quality implications, priority, time and duration, administrative procedures and effects on third parties. Water allocation rules specifically determine water deliveries, and their obligations therein, in an irrigation scheme.

[6] See Yan Tang 1991

[7] See Bottrall 1981, pp.122-127

Box 2.1. Water distribution methods[8]

- *On-demand:* Water is available to the farmer at any time when the intake or hydrant is opened. The amounts of water are not limited. This method requires high-level technology and it operates on automatic principles.
- *Semi-demand:* Water is made available within a few days of the farmer's request. The amount is often limited to a certain volume per hectare. The 'water guards' play a crucial role in facilitating communication amongst farmers, and the farmers and the higher level management units. The system shows a low technical efficiency in times of low demand, and considerable water losses.
- *Canal rotation and free demand:* Secondary channels receive water in turn, and once a channel has water farmers can take the amount they need at the time they wish.
- *Rotational system:* Secondary channels receive water in turn, and the individual farmers within a given channel area receive the water at a preset time, and, generally, in a limited quantity.
- *Continuous flow:* Throughout the irrigation season, the farmer receives a small but continuous flow that compensates the daily crop evapotranspiration.

The main distinction between the water distribution methods is whether they follow a demand-driven or a supply-driven concept.[9] The concepts have important impacts on an irrigation agency's planning and daily implementation efforts, and on supply. In *supply-driven concepts* the farmers have to adjust the crop decisions to water supply; usually less water is diverted than demanded. This solution is favoured where farms are abundant and small, because with demand-driven approaches planning and executing the many operations is costly, sometimes even impossible, and operations are poorly performed. With supply-driven concepts, the farmers have little say in the supply of water. It is reported that because there is little discretion, „collusion between head-end irrigators and canal officials is minimized. Most would acknowledge that such a system is sub-optimal. (Its advocates) argue, however, that the cost in crops not produced, because farmers have less freedom to optimize cropping, is more than offset by the fact that such a system is manageable, e.g., that predetermined water delivery schedules can be met with a high degree of certainty, so that farmers can bank on getting irrigation water at specific times and plan accordingly".[10] In *demand-driven, or crop-based, concepts*, the amount of irrigation water delivered is tailored to crops which the farmers choose; with this concept farmers are able to optimize their cropping, because the supply responds to individual demands. The semi-demand-driven solution is, perhaps, the most common, as Bottrall assumes. The supply decisions are determined by the availability of water, the supplier's perception of the users' reliability preferences, and by the technology and the state of the irrigation system. For planning purposes, the decision-makers need to gather paramount information: estimating the available water; estimating the water demands according to the crop pattern of a large number of farms where the farmers have a free choice of crop-

---

[8]   Source: Bottrall, in: FAO 1982, pp.65-69
[9]   The World Bank 1994, pp.103-106
[10]   The World Bank 1994, pp.104-105

ping (and the management unit has no authority over this issue). Estimating the water supply depends further on irrigation efficiencies at on-farm and on-project level, which are often not known. The daily implementation generates high monitoring and enforcement costs, and the management unit minimizes the costs by applying a uniform set of water deliveries that lack specificity to local conditions. One unit of water per hectare of irrigated land is released, and although the type of crops planted are sometimes taken into account, varying local conditions, such as soil permeability, the state of the irrigation network, farming practices, and application efficiencies at on-farm level are not. The uniformity can result in too much or too little water input, varying with local conditions. The decisions on the amount of water released per hectare according to the crops' water requirements are usually made by the irrigation agency, not by an agricultural service or the farmers. To achieve more flexibility, this task would have to shift from irrigation engineers to agriculturalists, and that would mean that the irrigation agency would have to give up power from within its domain of responsibility,[11] or it necessitates cooperation between engineers and agriculturalists, with high information and cooperation costs.[12]

In both concepts, the decisions for water or crops, respectively, are, however, conditioned by the irrigators' perception on the reliability of water supplies over which they have no control, their perception of the plants' water requirements, together with other factors of farm economy. Water is the essential input factor and affects other input variables. The farmers are dependent on receiving a specific quantity and quality at a particular time and location. These factors are equally important, as are the associations with the crops' water requirements, the land preparation, the harvesting cycle, and the available labour forces. Other inputs have to be adjusted to water deliveries, because if not, the farmers are wasting seed and fertilizer. Bromley emphasizes that „when decisions are made about the amount of land to be cultivated, or the particular crop to be planted, or the purchase of fertilizer, likely water availability and the probable timing of water receipts are crucial factors in (their) decisions".[13]

If the individual irrigators cannot be certain of the quantity, location and timing of the resource available, they respond with risk-minimizing strategies: they may select crops which do not have to be irrigated; they may secure the available water for certain selected crops and fields; they may plant crops which are more drought resistant; they may be discouraged from double cropping, or tend to use lower levels of inputs as a hedge against reduced input response during possible periods of water scarcity; they may spread scarce channel water over larger than optimum areas; they may over-irrigate their land, and they may oppose the officially imposed water allocation schedule by illegally withdrawing water.[14] The last possibility frequently occurs because the negative impacts on farm economy, i.e., reduced yields and low profits, can thus be prevented. Excess water may be applied,

---

[11] But even for an agricultural agency are information and coordination costs high.
[12] As Chambers (1988) states, calling for coordination is easily done, but it should be said what should be done, by whom, how, when and with what resources. (p.235)
[13] Bromley 1982, 2.32
[14] See Bromley et al. 1980; Wade 1975; Abel and Chambers 1976; Dudley 1992

which is of particular significance for the emergence of high groundwater levels and salinity. If inadequate services are the outcome of the operation activities, then the incentives on the side of the farmers for illegal water tapping increase for whatever reasons (their view of the plants' water requirements, farm operation conditions, insecurity of receiving no water, etc.), and withdrawing water in excess is a risk-minimizing option. Some empirical case studies provide evidence of farmers overirrigating their fields to reserve supplies of water in case of later delays in water delivery.[15] Mike Moore points out that in Taiwanese irrigation systems, which are known to be the world's most technically efficient ones, the rice farmers are tempted to steal water in excess of their needs:

> A deficit supply produces immediate and relatively severe yield decreases. By contrast, over a wide range, an excess supply does not harm the plant at all. It makes a great deal of sense for rice farmers to seize any opportunity to appropriate more water than they are likely to need. They can store it in their flooded fields for several days, and feel that they have provided for their crop if for some reason canal supplies prove deficient.[16]

This is, of course, only peculiar to paddy cultivation, because paddy is not sensitive to too much water, but it is well known and documented that farmers 'overirrigate' even when the planted crops are sensitive. If irrigators oppose the scheduling in time and quantity, they contribute towards higher groundwater tables and waterlogged lands, and water distribution problems occur downstream, where the tail-enders have no leverage over those at the head-end to release sufficient amounts.

In bureaucratic systems, the water allocation rules are usually intended to treat the irrigators' demands with equity, i.e., the rights of use appertain to landownership or lease, and no difference is made in favour of some farmers. But this aim is rarely achieved. Bromley, Taylor and Parker report:

> Our examination of the literature gives attention to the two main factors - farm location on the water course and farm size - that seem to underlie most cases of inequitable water distribution (...). First, farmers whose fields are most distant from the source of water frequently have the least secure water supplies. Their greater water insecurity is related to (a) the greater cumulative effect of seepage and evaporation losses in delivery channels as fields are more distant from the water source, and (b) the greater possibility for intervening irrigators to disturb intended water distribution as the water flows from head-end to tail-end fields. Second, farmers with larger holdings and other forms of economic power frequently have more secure water supplies.[17]

While the release time may be exceeded, though not very substantially, it may, however, not suit the farm's operating conditions, e.g., labour may not be available, and extra labourers have to be hired. The timing is crucial in relation to the opportunity costs of other farm activities. Many delivery systems are designed for diverting water 24 hours a day, and much water is badly used or wasted. Chambers addresses attention to this issue while reporting:

---

[15] Abel and Chambers (1976), in: Bromley et al. 1980, p.370
[16] Moore 1989, p.1741
[17] Bromley et al. 1980, pp.371-372

Darkness, cold, fear, normal working hours, and a desire for sleep deter irrigation staff, farmers and labourers from activities at night. At the farm level, irrigation at night entails extra labour and costs. It requires smaller streamflows and well-shaped fields. Paddy and trees are the easiest crops to irrigate, and younger, lower, and more thinly spread crops are usually easier than those which are older, taller and denser. Potential productivity of water at night is slightly raised by lower evaporation losses, but this gain is neglible compared with losses from breaks in channels, inefficient water application, and wasted water flowing into drains. Equity effects at night are mixed: some farmers poach at the expense of others, but some, who are denied it during the day, get water at night. Night irrigation increases costs and inconvenience to small farmers, but raises labourers' income.[18]

While some farmers may prefer night irrigation, for others the costs exceed possible benefits; whether the impacts of night irrigation are positive or negative depends on field conditions, crop types, and water reliability. Flooding and waterlogging can result from uncontrolled water flows at night, and water flowing uncontrolled through minors, watercourses, field channels and drains at night contributes to the problem of rising water tables.[19] Chambers evaluates means for water saving by reducing irrigation at night through storage in reservoirs, main channels, in intermediate storage reservoirs, and on-farm, but there are many difficulties and disadvantages, of which the most important is cost.[20] Night irrigation also poses similar inconveniences and increased costs on the irrigation agency, and field workers must be strongly motived to work at night.

In bureaucratic systems, the allocation of water is based on imposed water allocation rules, without the mutual consent of all irrigators. Although the management unit may try to execute the allocation rules impartially, and to create security in the water supply, on the side of the irrigators the temptation to object the allocation rules is present. The reasons for opposing the rules may be that the irrigators are not aware of the water availability, and, because it is beyond their control, they cannot be sure whether they will receive water; the imposed rules are inflexible in their response to local needs even within one service area; in times of peak demand water shortages occur; with general water shortages, demands exceed available water; farms at the head-end withdraw more water than conceded, because they have more power from position and location, and the individual irrigator cannot be sure that the other irrigators will follow the rules. Water stealing by an individual farmer, i.e., taking water at a time, and of a quantity, which does not coincide with the authorized rules, is, however, a 'rational' behaviour of the irrigators. If it occurs, then it may invoke a vicious circle where water stealing by one farmer may cause water stealing by others, because everybody tries to accomodate their own demands.

Rule enforcement creates institutional certainty for the irrigators, and is, therefore, regarded as a conditio-sine-qua-non for the efficient allocation of a common-pool resource.[21] Rule enforcement is costly, and needs effective institutions to limit

---

[18] Chambers 1988, p.133
[19] Ibid., p.147
[20] See Chambers (1988) who provides a very detailed and informative discription and analysis (pp.148-153)
[21] See Ostrom 1990, 1992; Weissing and Ostrom 1993; Yan Tang 1991; Hilton 1992

transaction costs. Rule enforcement in government-organized systems is influenced by several factors, for example, the range of punishments assigned to rule breakers, the probability of detection, and whether sanctions are executed or not. Punishments range from simple warnings and fines, to water cut-off. The threat of punishment only affects behaviour if rule opposers face the threat of being detected, and of sanctions being executed. Monitoring the irrigators and detecting the rule breakers is, therefore, an issue of particular importance. Weissing and Ostrom oppose the predominant assumption in the social science literature which presumes that „increasing the fines imposed on those that steal is the most important parameter".[22] They conclude that the incentives for the potential rule breaker, i.e., the imposed punishments, have little or no influence on rule-breaking rates. The incentives for guards or irrigators to monitor and detect are regarded to be more important. An equilibrium then is present where stealing is less than maximal.

In most government-managed irrigation systems, 'disassociated' guards, i.e., government employed labourers without intrinsic interests, have to ensure that the rules are enforced. In systems with disassociated guards, the guard's salary is not related to the overall performance of the system, and they do not suffer losses due to undetected stealing events, nor are they affected if a stealing attempt is detected by an irrigator and not by them.[23] The guards are not interested per se in a reduction of the stealing rate, but if they are rewarded, and if this reward is high enough in comparison to the costs of guarding, the stealing rate is reduced. The variables which influence the incentives for the guards, are the rewards assigned for successfully detecting stealing, the costs of monitoring and the losses suffered if stealing is not detected.

In bureaucratic systems, it is assumed that there are counter-incentives and restrictions on monitoring and detecting: the guards are employed as labourers by the state agency, and are supervised by its engineers, not the irrigators. They might prefer ease of work, and the edge is taken off of internal control. The supervising engineers might not gain professional or other rewards. Public officials might not want to behave like policemen; they might have no incentives to be actively involved in solving farmers' problems. The execution of sanctions might not be desired, politically and/or socially, or the agency might not be allowed to execute punishments, thus monitoring and detecting efforts become useless. Public officials do not identify themselves with the local systems. They belong to a thinly manned O&M department with inadequate transport facilities, which sets constraints on monitoring; their numbers might be insufficient in relation with the service area. Their working hours might not suit those of the system's operation, particularly if night irrigation is practiced. The farmers' refrain from the repayment of water charges will have no effect on the guards' salary, and the farmers have no influence over the performance of the field guards or their supervisors.[24] Wade attributes some of these counterincentives to a more structural problem which has affects on the guards and their supervisors: there is no connection between the budget and the collected water charges, and no linkage between the allocated

---

[22] Weissing and Ostrom 1993, p.425
[23] See Weissing and Ostrom 1993
[24] See Yan Tang 1991; Weissing and Ostrom 1993; Wade 1987; Chambers 1988

budget and the performance of the schemes; the operation activities are carried out by organizations whose primary function has been construction.[25]

The incentive structures facing public officials have been evaluated by Elinor Ostrom, drawing on a study by Hilton who analyzed irrigation systems in Nepal.[26] The features which affect the behaviour pattern of irrigation officials are, first, that for recruitment and promotion. The ability of an official to work with farmers or to solve the day-to-day problems of an irrigation systems plays no role; second, promotion is rendered ineffective as a motivational tool by the length of time it takes; third, seniority is the most important criterion for promotion, and individual initiative and creativity are discouraged, and irrigation officials who do not commit serious mistakes or offend their superiors will eventually be promoted; fourth, corruption is part of the 'day-to-day' business, and given the low salaries of public officials, counteracting accepted bribes is difficult; fifth, officials within the O&M departments, which anyway lack sufficient funds, have a low status, and finally, civil servants prefer to live in urban areas. Ostrom concludes that there is a complete lack of any intrinsic or extrinsic rewards to government officials for keeping an irrigation system in good condition.[27]

The agency's involvement and its effectiveness in solving conflict over water does not only depend on the willingness of the public officials and on rewards,[28] but on formal or informal societal conflict-settlement arrangements, which differ among societies, and the legal system has important implications for conflict resolution in irrigation systems.[29] The agency's power may be restricted, and public officials may have limited possibilities for intervening. If conflicts such as farmer-farmer, or farmer-agency disputes have to be settled by a court, be it whether the state agency or the farmers are the plaintiffs, the transaction costs may offset the benefits for both parties.[30]

---

[25] Wade 1987, p.179

[26] Ostrom 1994; Hilton 1990, 1994

[27] Ostrom 1994, pp.211-212

[28] Their willingness or unwillingness may also depend on 'extra' contributions from some farmers.

[29] Katherine S. Newman's (1983) typology of legal institutions can serve as a means to characterize conflict-settlement institutions in irrigation projects although she tries to characterize pre-industrial societies by types of legal institutions. She identifies several dimensions: The first dimension is the presence or absence of a third (non-disputant) party, or parties in the settlement process; the second concerns the extent to which disputing parties are socially required to turn to this third party for dispute settlement. The third dimension concerns the intended authoritativeness of third-party decisions, the fourth relates to the degree of centralization or concentration of legal decision-making power in a third party, and the last dimension concerns the issue of levels of appeal, or levels of jurisdiction.

[30] The farmers are, however, in a weak position because they usually have no distinct and clearly defined water rights.

1.2
Maintenance Versus New Investments

The maintenance of irrigation and drainage systems can secure high water convey-ance efficiencies, reducing seepage and percolation, and guarantees that excess water will be drained out from the area, and is, thus, an important means for groundwater level and salinity control. Carruthers and Morrison define certain types of the maintenance activities, such as normal or routine, emergency, de-ferred, catch-up, preventative, condition-based and imposed maintenance. Mainte-nance categories consist of desilting, weed control, the maintenance of regulatory structures, and of the mechanical equipment.[31] The extent and the kind of mainte-nance requirements depend on the design and quality of construction which is poor in many projects.[32] If high investment levels are employed, for example, concrete-lined channels instead of earth-lined, the mechanical and manual items change, and with them the costs for maintenance. The benefits from lining can be assumed to be the value of water saved by the reduction in seepage losses. Seepage rates can be reduced by 20% of the unlined value. Concrete lining has the advantage of reduced maintenance activities, but the disadvantage of increased maintenance costs, e.g., the costs for repairing cracks might be high.[33]

Karunasena regards maintenance as a „management response to the deteriora-tion of the physical condition of irrigation systems that threatens to make it im-possible to achieve operational targets".[34] Maintenance and its financing is a key determinant of the long-term viability of irrigation and drainage systems. Ostrom et al. emphasize that because „it can increase the productivity of capital and/or prolong its useful life, maintenance itself can be regarded as an investment".[35] Carruthers and Morrison point to the dominant weight of original capital as sunk costs in irrigation projects, and conclude that „the returns to investment in mainte-nance are likely to be very high. Maintenance investment is good economics".[36] Although this is acknowledged, maintenance is a neglected issue, and this has occurred most markedly in public sector projects. Maintenance is a much more serious problem than statistical data could demonstrate or suggest, and a much more striking feature is the neglected maintenance of drainage systems.[37] Since high groundwater levels and waterlogging may take some time to develop, the magnitude of the problems may not be realized until they become very serious. The increasing rehabilitation needs of many public schemes can serve as an indica-tor for deferred maintenance. Maintenance activities have not kept track with the expanding irrigated areas, and the rehabilitation needed to overcome widespread construction deficiencies and neglected maintenance overwhelm national budg-

---

[31] Carruthers and Morrison 1994, pp.21-22; see Bottrall 1981
[32] See The World Bank 1994
[33] See Goldsmith and Makin 1989
[34] Karunasena (1993), quoted in Carruthers and Morrison 1994, p.14
[35] Ostrom et al. 1993, p.34
[36] Carruthers and Morrison 1994, p.17
[37] See Hotes 1983; Repetto 1986; Biswas 1990

ets.[38] Scarce financial resources may be allocated to other uses (viewed as higher-value), and only a small amount is allocated to maintenance activities. New schemes are constructed, according to a state's needs, at the expense of maintaining existing ones, and „once built, many agencies are not able to perform the necessary operations and maintenance".[39] The strong trend for new installations is described in a World Bank study:

> Worldwide, there seems to be a deep-seated human tendency to want to do something new and to denigrate what is mundane and day-to-day. The exitement of inaugurating a new project is greater than that of maintaining existing ones. In irrigation, new construction seems to get unwarranted precedence over not only operation and maintenance, but also over the up-grading of existing schemes, and even over completing new schemes that have been started but not completed. Cutting the ribbon for a new project gets more publicity than routine O&M, even though the latter may produce more benefits at less costs.[40]

If political decision-makers can gain access to external funds with the construction of new projects, scarce local resources are allocated to new investments, with maintenance being omitted. In many countries, increasing the supply of water through new investments is politically more convenient. The decisions to develop new water sources to extend the supply may be supported by the water users, and may even be a result of their pressure on policies. Repetto assumes that the pressure to develop new water resources is „likely to be greater if the water users do not expect to pay for the costs of investment, and could lead government agencies to make uneconomic decisions".[41] Small and Carruthers consider that the water users' interests may coincide with professional engineers and other specialists in an irrigation planning and design department who are trained to plan and design projects. These specialists „therefore have a vested interest in seeing that new projects are brought forth for planning and implementation. The ultimate economic performance of these projects is of little immediate importance to them" as this issue is the responsibility of irrigation engineers who, on the other hand, have not been involved in planning and design. In the case of poor performance, „a burden on the general economy (...) is largely hidden and has no direct financial implications for the irrigation agency" which receives its investment budgets independent of the actual performance.[42] The authors point out that „under central financing, the amount of money available to an irrigation agency for investment in new projects depends on the agency's ability to convince higher levels of government of the desirability of the investments", and this ability has nothing to do with the actual performance of the projects.[43]

---

[38] Frederiksen 1992, p.3
[39] The World Bank 1993, p.101
[40] The World Bank 1994, p.29; see Edelman 1977
[41] Repetto (1986, p.24) quoting from Small et al. 1986, p.11; see also Small and Carruthers 1991, p.100
[42] Small and Carruthers 1991, p.98
[43] Ibid., p.98

It is commonly agreed that the most widespread cause for poor maintenance is the lack of sufficient funds to undertake maintenance activities adequately,[44] and the classic remedy for maintenance problems has been to cover the costs with the water charges.

> This issue is extremely important since the capacity of the organization to carry out work efficiently depends on it. The question of insufficient funds is a rather complex one because many factors - social, organizational, economic and political - interact, and it is often difficult to determine the real origin of the problem and how to break the existing vicious circle. All the expenditure for running the water management organization should in theory be covered by the water rates, and the Administrative Service should control that income and expenditure are in equilibrium. However, there is often a large gap between the funds collected from this source and the actual expenditure on the scheme. This gap is sometimes bridged with a subsidy from the government, particularly in public irrigation schemes, or more commonly by not undertaking the necessary maintenance.[45]

Carruthers and Clark underline the consequences of insufficient budgets for the O&M departments: Once starved of funds and equipment, they became low prestige areas and cannot attract skilled personnel.[46] Apart from the insufficient funds available to irrigation agencies, which may derive from the inability or unwillingness of governments, and from the lack of their strategic vision to allocate adequate resources, poor work organization and lack of interest from public officials are other causes for the demise.[47] In performing the maintenance tasks, the agencies may be unable to secure information about local maintenance requirements. Establishing adequate maintenance services, (e.g., the frequency of maintenance servicing for each kilometer together with the equipment required) one must take into account the importance of each drain and delivery channel, the type of weeds growing, and the volume of silt to be excavated.[48] Usually, routine annual maintenance programmes are set which might not suit the real local maintenance needs, and which do not respond to field realities, because the agency cannot achieve these high information requirements.

Elinor Ostrom et al. have formulated a set of research hypotheses to explain the systemic failures of maintenance decisions in public irrigation projects.[49] They assume that the factors which influence maintenance in public irrigation schemes are more complex than in private enterprises. There, the basic economic model of investment suggests that maintenance will be performed only if it yields a rate of return greater than that resulting from alternative uses of the resources so invested. Two factors are important in public systems: firstly, public irrigation infrastructures are used in common by a large number of inviduals, and, secondly, there is more than one decision-maker. The non-excludability character of the generated benefits provide incentives for the individual user not to contribute to the provision

---

[44] See Sagardoy et al. 1982; Biswas 1990; Carruthers and Clark 1981
[45] Sagardoy et al. 1982, p.130
[46] Carruthers and Clark 1981
[47] See Sagardoy et al. 1982
[48] See ICID 1969
[49] Ostrom et al. 1993

of the goods. It is costly to exclude non-contributors, and even if public officials have the authority to impose water charges and to regulate the use of the goods, it often appears that maintenance is subsidized by governments. The benefits of maintenance may not immediately be visible, and if the rate at which future net benefits are discounted is high, maintenance decisions may be neglected. One may also add that in the case of poorly maintained public drainage systems, the damages and costs on farm economy might only appear a couple of years later, after the operation has started. The public officials do not, however, bear the costs of neglected maintenance. Although all irrigators are polluters - all contribute to high groundwater levels - not all irrigators are victims of poorly maintained drainage systems, and the need for investments and better maintenance might be perceived only by some affected irrigators. Then, demand for drainage is weak, and collective-action problems arise. If maintenance is neglected, be it the maintenance of the delivery or the drainage systems, cumulative effects occur with breakdowns in the system's effective capacity, and high replacement costs can emerge.[50] If only a few contribute to maintenance, but all can benefit, then the problem will be to arrange a system of incentives to reduce free-riding. The multiplicity of decision-makers is the second important factor influencing maintenance decisions in public irrigation schemes: the legislature, government officials, public officials in one or more ministries, local administrative unit officials, and the users. Costs and benefits are differently perceived and assessed, and thus future costs/benefits of reduced rates of capital deterioriation are difficult to measure. Public officials must be strongly motivated if they are to invest scarce resources in efforts to reduce an imperceptible rate of deterioriation, rather than for activities producing more obvious and immediate returns. Sufficient resources must be available together with a willingness on the part of the decision-makers to allocate funds for that purpose. It might seem to be more 'profitable' for government officials and the ministerial public officials to invest in new projects which show immediate financial returns, or which have higher support from their political clientel.[51] Wade, Skutcsh, and Ostrom et al. attribute neglected maintenance to the lack of an incentive structure in government agencies, and stress the need for motivational and incentive changes to develop a maintenance culture.[52] Maintenance activities rarely meet the needs of the irrigators who have, however, no direct control over the maintenance of the system; although they might be better informed on the local requirements, cooperation between the state agency and the irrigators has remained poor.

1.3
Self-Financing Principles for Irrigation Services in Public Irrigation Projects and Their Relation to Performance

The lack of sufficient funds in public irrigation schemes is reported to be the most widespread cause for inadequate operation and maintenance. The O&M divisions

---

[50] The particular problems which drainage systems create, are not considered by Ostrom et al.

[51] See Ostrom et al. 1993

[52] Wade 1993; Skutcsh 1993; Ostrom et al. 1993

are poorly equipped, and have too little funds for the employment of sufficient staff and for their adequate training. The recovery of recurrent costs through water charges, however, is regarded as a financial means to overcome this constraint. The total incomes for O&M should be in an equilibrium with the expenditures, and the water fees charged to the farmers should be set at levels required to fully and adequately maintain and operate the facilities. Since the 1980s, the World Bank has placed high emphasis on cost recovery for O&M expenditures in Bank supported irrigation projects, with some contribution, also, to investment costs, because „those who benefit from public irrigation projects should pay for what they get".[53]

The classical remedy of recurrent cost recovery through water charges has shown little positive effects. As reviews show, in most developing countries, water charges are assessed at levels lower than the O&M costs incurred, and „yet, even these costs, which are typically only 10% or less of the total costs of major surface water irrigation systems, are rarely covered".[54] The egyptian farmers, for example, have no obligations; in Jordan and in Marocco, but also in Australia and Canada, the farmers pay less than the production costs, and only in France do the farmers pay 50% of the costs incurred.[55] Cummings and Nercissiantz, who compared irrigation projects in the United States and Mexico, found that in the US projects, the farmers pay the full amount of O&M costs which are assessed at levels required to fully maintain and operate the facilities. In contrast, the Mexican farmers pay only a small proportion (15 to 25%) of O&M expenditures, which is less than required to adequately maintain the irrigation facilities.[56] Even in countries where legal regulations determine that O&M costs should be totally recovered by water charges, the collection rate never reaches this target. The rate of water charge collection depends on the administrative system of control, on the system of collection, and on the penalities imposed and enforced on non-payers, like the termination of irrigation services, fines, etc. The enforcement of sanctions to non-payers imposes high costs, and is further dependent on the technology of the delivery system (if water-cut offs are considered), and if the ministries of finance are responsible to administer and collect the charges, cooperation with the irrigation agency that has to enforce penalties is necessary.

The collection of water charges from a large number of irrigators, however, requires expenditures to administer and enforce the collection. Water charges will contribute little to the net resources of the finance ministries if the collection costs are high. Small points out that the effectiveness of finance methods cannot be truly investigated if collection costs are not considered; but data on the costs of collection in relation to the collected water charges are rare.[57] Small and Carruthers demand 'active collection efforts' as an important element of the government's enforcement procedures instead of „passive procedures, such as waiting for farmers

---

[53] Jones 1994, p.41; see The World Bank 1994
[54] Repetto 1986, p.19; see FAO 1982; Small et al. 1989; Schiffler 1993
[55] See Tsur and Dinar 1995, pp.4-8
[56] Cummings and Nercissiantz 1992
[57] Small 1989

to come to the central office to pay their fees".[58] In general, inherent incentives are regarded to be necessary to make the farmers pay their fees, which is not the case when funds are provided for services irrespective of the repayments,[59] and the authors consider incentives for the state agency to take positive actions to obtain high rates of fee collection.[60]

Even if one supposes that the charges would be at levels set to recover the costs, and that they could be collected from the beneficiaries, they are paid to the general treasury, and it depends on political decisions as to whether the budgets for O&M are sufficient. However, on the public irrigation projects in a large number of countries, the amount of water charges collected from farmers bears no relation to the amount of funds which are made available for O&M, especially if the operating agencies depend solely on governmental allocations for its revenues.[61] Financially dependent irrigation agencies have usually neither the legal responsibility to assess water charges nor to collect them, and they have no means of control over the beneficiaries' payments. The staff in O&M divisions is accountable to the central administration, but not to the farmers. The farmers have no means of pressing for good services, and their unwillingness to pay and the withholding of payments have no effects on fund provision and services. Sagardoy et al. paraphrase these non-linkages, where there is no feedback at any level as follows:

> On government-run projects, there is usually no direct link between the level of performance a project organization achieves and its financial rewards. The amount of funds it receives from government is determined first by the total availability of funds for recurrent expenditures within the irrigation and agriculture sectors and then by reference to „yardsticks" which reflect the government's view as to the relative operation and maintenance needs of different projects. The quality of the project's performance, either in terms of increased agricultural production or in its recovery of revenue through water charges or other forms of taxation, is not a criterion of how much finance it receives. Such a system removes any incentive for farmers to pay their water charges, since the proceeds go to general revenue and have no direct effect on the level of local reinvestment. There is similarly no material incentive to management and staff to improve their service to farmers since it will not lead to any increase in their financial rewards. Further side-effects of this policy are a hardening of farmers' resolution to oppose attempts to raise water charges; a reluctance to pay water charges and other taxes, even at the very low rates often prevailing; and a consequent need for revenue-collecting staff to be more coercive in their relations with farmers.[62]

On the other hand, it is reasonable to assume that even with larger budgetary outlays to irrigation agencies from the central treasury improvements are not automatically forthcoming, and there is „ample evidence to suggest that O&M done by public irrigation agencies is often neither cheap nor effective".[63]

---

[58] Small and Carruthers 1991, p.110
[59] See Small and Carruthers, who dedicated a chapter to "Collecting irrigation fees: Fostering a willingness to pay" (1991, pp.182-202).
[60] See Small and Carruthers 1991, p.202
[61] See Small and Carruthers 1991, pp.45-46
[62] Sagardoy et al. 1982, pp.53-54
[63] The World Bank 1994, p.88

1.4
Economic Means for Achieving Efficient Water Use and for Controlling High Groundwater Levels and Salinization

The effects of O&M cost-bearing on water-use efficiency are, however, limited because water usually is charged on an area basis, and this does not provide incentives for water savings. On the contrary, this method can promote the wasteful use of water because in the farmers' decision-making the area-based charges are fixed costs of production, regardless of the actual water-use decisions. The magnitude of the water charges will have no effect on the quantity of water used. Imposing water charges on an area basis is a widespread phenomenon: Bos and Walters studied farmers representing 12.2 million ha of irrigated farms worldwide, and they found that in more than 60% of the cases, water charges are levied as flat rates.[64] If water is scarce, and wasted, recurrent cost recovery for O&M alone will not suffice to instigate efficient water use, and the incentives for water saving must be stronger than devolving O&M costs to the users.

Recognizing the increasing scarcity of water, it is proposed that water-use efficiency should be promoted through water prices which reflect its opportunity costs. The opportunity costs of water provides a measure of the scarcity-value of water to society so that, taking into account society's multiple objectives and water's multiple uses and interdependencies, any cross-sectoral differences in value are highlighted. Information about, and the analysis of, supply options, future demand, investment alternatives, together with the economic costs of pollution and other environmental damages are required when determining the opportunity costs of water.[65] Farmers should be discouraged from using too much water by increased water charges. It is assumed that this would instigate water savings, which is desirable from an environmental point of view, because water savings in irrigated agriculture contribute to reduced high groundwater levels, waterlogging and salinity, and limit, on the other hand, the demand for increased water supplies. The options available to private farm-firms against higher water prices depend on the relation of costs and expected net benefits, and they may influence the farmers' decisions in different ways (short- and long-term responses): the amount of water to a given crop can be reduced with or without water-saving technologies, with little costs (i.e., yield losses); other crops are planted which do not need irrigation water, or another cropping pattern can be applied with reduced water requirements, etc.

Raising water charges, however, can serve as an allocative device only if they are based on the volume of water used, or other proxies.[66] The on-demand water delivery method poses no constraints on the volume delivered, the semi-demand system, however, delivers water at the request of the farmers, and the amount is limited to a certain volume per hectare. The water charges under these systems are a function of area, crop, season, application method and/or number of irrigations, and do not reflect the volume. Under the rotational system, which delivers water in

---

[64] Bos and Walters 1990, in: Tsur and Dinar 1995, p.4
[65] The World Bank 1993, pp.42-43
[66] Chambers states that "this is not the basis anywhere in Asia, including Taiwan where water management is relatively sophisticated and efficient". (1988, p.83)

turns according to a predetermined schedule, the water is annually charged on the number of shares, or the proportion of water received; this charging system shows, at least, some relation to the applied volume.[67] Volumetric pricing is generally regarded as a means for improving water-use efficiency. Leslie E. Small points to the effects of the not very common volumetric pricing system which depends on the farmers' range of discretion. He has evaluated volumetric pricing systems in Northern African countries stating that „a distinction has to be made between situations where (...) the farmer has little or no control over the volume of water received, and those situations where individual farmers make decisions on the volume of water to receive. The latter cases represent situations of true water pricing (..). The former situation amounts to a special form of an irrigation service fee similar to a flat charge for water per unit of land area".[68] Small assumes that „the incentive for farmers to be efficient in the use of water is provided by the water allocation mechanism, indepedently of the method of financing irrigation services".[69]

An internalization of unwanted external effects generated by agricultural drainage effluents can be reached through state regulations, and in public irrigation schemes the common means are that the farmers contribute towards investment costs and the cost for drainage system maintenance; the individuals' contributions are, however, assessed on an area basis, irrespective of the applied amount of water, or the drainage effluents. Pollution taxes could be applied which requires regulation standards for effluents as a precondition; but monitoring and measurement is, however, difficult, and collecting the taxes would face the same problems as the collection of O&M water charges. The extent to which an individual irrigator benefits from and uses the drainage systems (i.e., the emissions from individual farms), cannot be technically measured or, at least, with extreme difficulties and expense, and would show high degrees of uncertainty due to varying physical conditions, such as soil permeability, water input and irrigation practices. The number of polluters and victims are large, with the two groups often overlapping. Those adversely affected cannot always be separated spatially from the pollutant sources within an irrigation project area. Measurement problems occur, furthermore, because the effects of inadequate drainage systems vary. High groundwater levels and salinity have to be regarded as distinct from each other, because they affect the crops in different ways. These two factors are, however, interrelated. The emergence of high groundwater levels and salinity depends on a large number of physical conditions all of which may differ locally, such as soil permeability, type of crops, irrigation practices, climate conditions, water input, etc. Measuring the contribution of n-farmers to rising groundwater levels, or measuring the salt discharge from each field is impractical. The contributions from individual economic agents cannot practically be measured by direct monitoring.[70] Therefore, output-oriented policies are extended to inputs by postulating that externality levels may be dependant on the amount of productive factors employed, and can be

---

[67] OECD 1987, p.91
[68] Small 1989, p.10
[69] Ibid., p.16
[70] Griffin and Bromley 1982, p.548

applied to every factor on which the externality generation depends.[71] Factors influencing emissions can be measured at more reliably cost, despite the fact that unresolved problems remain, e.g., varying soil permeability, topographic conditions; the necessary condition for measuring water-inputs is, however, usually not given.

Groundwater and salinization control can be encouraged with water-saving (or, input-saving) technologies with lower pollution coefficients; it has been revealed, however, that farmers choose these technologies only in combination with higher value crops which yield higher benefits. Schiffler, for example, evaluated water-saving options in Jordan; he found that only with the increase in profits from higher value crops are the technical innovations viable.[72] In the cotton production of the Western Sao Joaquin Valley (California), Caswell et al. who modelled the effects of pricing policies on water conservation and drainage for irrigation districts in the United States, found that there is little variation in yield by the adaption to modern technology, but drainage per hectare could be reduced significantly by up to 85% with a switch from furrow to drip, and by 11% to modified furrow technology.[73] This shift to water-saving technologies has not been induced by higher water prices but by subsidies.

In large-scale public irrigation systems in most developing, and developed, countries, the water charges rarely function as an allocative device; they do not even cover the production costs. Pollution taxes are rarely applied, and the most common approach to agricultural effluents is that the assessed charges should recover the maintenance costs of the drainage systems. But the rates of repayments are low having no influence on budget allocations or on the quality of the supplied maintenance services.

2.
Small-Scale Farmer-Owned and -Managed Irrigation Schemes

The poor performance and failure of large-scale public irrigation projects explains the increasing interest in reviving or creating new farmer-owned irrigation systems to manage common-pool resources. In the economics literature, these systems are labelled 'irrigated common-property regimes'.[74] Empirical research has concentrated on traditional irrigated common-property regimes, and they are supposed to make long-range sustainable use of the water resource and the resource systems; they are self-financed by the direct beneficiaries through local resources; they are able to limit free-rider behaviour, rent-seeking and corruption. In most systems, scarce water is allocated efficiently. Successful communal systems are small in

---

[71] Meade (1952), quoted in Griffin and Bromley 1982, p.549

[72] See Schiffler 1993, pp.48; 50
Low-cost technologies are not always available for all type of crops.

[73] see Caswell et al. 1990; Gardner and Young 1988

[74] In this study the terms 'irrigated common-property regime' and 'farmer-owned and -managed irrigation system' are used synonymously.

size, and have notions about how large they could be, and how many units they could manageably contain. In reviewed World Bank-sponsored irrigation projects, the sizes range from 8 ha in India to 1,000 ha in Madagaskar, with an average size of 40 ha. But in „some Latin American countries it was up to several thousand ha, because the average family farm size is in the tens (or sometimes hundreds) of ha".[75] A clear figure cannot be given regarding the number of members, but as Cernea and Meinzen-Dick emphasize „the costs of maintaining an organization, particularly in terms of conflict resolution and information management, will increase with the size of an organization".[76]

Irrigated common-property regimes are reviewed here as alternatives to public irrigation schemes in a two-fold sense: first, as an alternative institutional arrangement where a common-pool resource is owned and managed as the private property of a user group, and, second, in that self-governed groups operate and maintain the lower levels in public systems, and successful incentive-reward mechanisms might be introduced into government-managed schemes. My particular interest concerns the potentials of irrigated common-property regimes for groundwater and salinity control. This faces some difficulties, however, because the reviewed empirical studies do not deal with this subject, either because of favourable natural conditions, the small size of the schemes, or because this problem was beyond the inquiry point of the scholars. Research on irrigated common-property regimes concentrates on water allocation in combination with the maintenance of the delivery systems, and, in general, on the conditions which guarantee the stability of common-property regimes. The points relevant to this discussion are how decisions on collective actions are forthcoming for the provision and the sustainment of common goods, how the interdependencies generated by a common-pool resource system are mediated through the group, and how water-use efficiency is achieved.

2.1
Common-Property Regimes by Definition, and Design Principles for Self-Governing Irrigation Institutions

Until recently, the term 'common-property' was synonymously used for resource systems where there are no recognized rights, and where everybody has free access. This mislabelling derived from The Tragedy of the Commons from Garrett Hardin who assumed that all situations in which individuals jointly use a resource would inevitably lead to the resource's deterioriation.[77] It was thought that individual rationality leads to outcomes that are not rational from the perspective of the group.[78] But many jointly used commons do not necessarily generate common dilemma situations. The term 'common-property', as used in this study, refers to

---

[75] Cernea and Meinzen-Dick 1992, p.49
[76] Ibid., p.50
[77] G.Hardin 1968
[78] Other social sciences' approaches support Hardin's analysis, e.g., the Prisoners' Dilemma and the problems of collective action (Buchanan, Olson). They all share the pessimism that individual rationality leads to 'collective self-damage' or harm.

the private property of a group exclusively owned by its members who have the right to exclude non-owners, and where the non-owners have to abide by this exclusion.[79] Or, as it is defined by Ciriacy-Wantrup, as the „distribution of property rights in resources in which a number of owners are co-equal in their rights to use the resource".[80] Common-property regimes are to be distinguished from free-access, because when access is free for all, there is no property at all: non-owned commons are not synonymous with common-property.

In common-property regimes, the members eligible to use the commons and the internal rules of use are defined. These rules, which govern the users of the resource, „explicitly distinguish the use of irrigation water from open-access conditions". They „cover acquisition, allocation and distribution, and operation and maintenance of the irrigation system. The collective performance of these activities necessitates the enforcement of formal or informal rules which define the individual water user's rights to receive a reliable water supply".[81] However, the interdependencies between the users are mediated through the groups' decision process, which is described by Bromley:

> A political forum is the nexus of interdependence; it is the place where conflicts over scarce resources are heard and settled. It is an interdependence that is both recognized and confronted. The users are the group over which costs and benefits are distributed - that is, costs and benefits of joint use are internalized to the group.[82]

This internalization in common-property regimes is unlike that in private property of inviduals, where interdependencies are not mediated by the direct participants, but instead are external to the decision calculus of any economic agent, and the internalization is unlike that in public schemes, where interdependencies are mediated by governmental agencies.

There is a great variety in farmer-owned and -managed irrigation systems, but, according to their institutional design, they have some principles in common: those who decide on the provision of common goods are those who benefit from their use; those who contribute to the maintenance of the systems, are those who define the operational rules. The initial provision and recurrent costs are fully recovered by the beneficiaries, and the individual contributions define the share of user rights. Decisions are either made by all members, or, in multilayer organizations, by their representatives; organizational activities, like daily decision-making, communication, conflict management, and resource mobilization, may be executed either by elected officials, or representatives chosen from among the farmers, by village elders, or hereditary chieftains. The members themselves may be directly involved in technical matters, or personnel may be employed by the group.[83] Generally, it can be said that the beneficiaries have means of influence and control, and they govern themselves through collective-choice arrangements. The most

---

[79] See Bromley 1989; Ostrom 1990; McKean 1992

[80] Ciriacy-Wantrup is of the opinion that "economist are not free to use the concept under conditions where no institutional arrangement exists. Common property is not everybody's property". (1975, p.26)

[81] Cruz 1989, p.221

[82] Bromley 1989, p.207

[83] See Cernea and Meinzen-Dick 1992, pp.51

elaborated design principles for long-enduring and self-organized systems are laid down by Elinor Ostrom (see Box 2.2.).[84]

Box 2.2. Elinor Ostrom's design principles for irrigated common-property regimes[85]

1. Clearly defined boundaries
   Both the boundaries of the service area and the individuals or households with rights to use water from an irrigation system are clearly defined.
2. Proportional equivalence between benefits and costs
   Rules specifying the amount of water that an irrigator is allocated are related to local conditions and to rules requiring labour, materials, and/or money inputs.
3. Collective-choice arrangements
   Most individuals affected by the operational rules are included in the group that can modify these rules.
4. Monitoring
   Monitors, who actively audit physical conditions and irrigators' behavior, are accountable to the users and/or are the users themselves.
5. Graduated sanctions
   Users who violate operational rules are likely to receive graduated sanctions (depending on the seriousness and context of the offense) from other users, from officials accountable to these users, or both.
6. Conflict-resolution mechanisms
   Users and their officials have rapid access to low-cost local arenas to resolve conflicts between the users or between users and officials.
7. Minimal recognition of rights to organize
   The rights of users to devise their own institutions are not challenged by external government authorities.
8. Nested enterprises
   Appropriation, provision, monitoring, enforcement, conflict resolution, and governance activities are organized in multiple layers of nested enterprises."

---

[84] Ostrom 1992, pp.69-76. See also Gibbs and Bromley who distinguish well-functioning common-property regimes by "(a) a minimum (or absence) of disputes and limited effort necessary to maintain compliance: the regime will be efficient; (b) a capacity to cope with progressive changes through adaption, such as the arrival of new production techniques: the regime will be stable; (c) a capacity to accomodate surprise or sudden shocks: the regime will be resilient; and (d) a shared perception of fairness among the members with respect to inputs and outcomes: the regime will be equitable". (1989, p.26)

[85] Source: Ostrom 1990, p.90

2.2
Collective Action for the Provision of Common Goods

The decision on the provision of a common good is by no means automatically forthcoming, even when all or most cultivators in a village or all potential users of a common-pool resource system could benefit from joint action. Preferences amongst the users have to be discussed, and a common agreement must be reached. It is assumed that underinvestment is more likely to occur than over-investment. The reviewed case studies assume that similar economic interests of the eligible users, such as land size and crops, and similar objectives in farm eco-nomics (subsistence or market oriented) promote collective actions. Wade, gaining experiences from Indian village irrigation systems, emphasizes that collective actions are likely to occur when there is a high risk without collective action, when the expected net benefit exceeds the transaction costs, when there is an urgent need (survival) to act for more than one individual, and when no other individual strategies are possible, or only such which promise less benefit.[86] The Zanjeras in the Philippines, for example, were formed through the initiative of a group of in-dividuals who offered to construct irrigation structures in exchange for the right to cultivate some land. This right could not be achieved without collective action.[87] The beneficiaries decide on the shares each member has to contribute towards the provision of a resource system, whether in cash, in material, or in labour, and, as a general principle, the benefits which the eligible members gain from the common goods show proportionate relations to this initial contribution. In a traditional small-scale irrigation scheme in Somalia, for example, the extent of the individu-als' involvement in building the canals is in proportion to the size of their land. The performance expected from the individuals is laid down by the size of land to be annexed. Should the labour requirements exceed the capacity of some farms' labour force, then these farms have to hire workers. The costs of construction which are not offset by the labour of the individuals themselves are covered col-lectively by all who will profit from the canal. Individual contributions are moni-tored and collected.[88]

2.3
Water User Rights, Water Allocation Rules, and Operation Activities

In common-property regimes, the members eligible to use the commons are de-fined. McKean, studying common rights to land and water in Japan, reports that „precisely who was eligible varied: all residents of a village, all taxpaying resi-dents of a village, all households that paid regular dues to the Shinto shrine, all households headed by an able-bodied adult male (...), all households with cultiva-

---

[86] Wade 1988, p.188
  These conditions are consistent with Olson's who assumed that smaller groups can easier agree, and if there are strong incentives only for some of the individuals.
[87] See Cruz 1989, pp.218-235; Ostrom 1990, pp.82-88
[88] See Boguslawski 1986; Seger 1986

tion rights, and so on".[89] In irrigated common-property regimes, the rights to water might be inhered in land or separated from land, they might refer to a time-share or a quantity, and they are only bestowed on the members of a resource community. In a Northern district of Pakistan, water rights are associated with land tenure, but in areas of acute water shortage, water rights and land rights are quite distinct from one another, and transactions for water shares take place in some of the villages.[90] However, the eligible members have to perform their full duties and the group exerts strong pressure on free-riders. The user rights may be obtained by contribution to the initial provision of a resource system; the user rights might only remain with the descendants of the original builders, or shares might be purchased or leased. Furthermore, a continous right can only be upheld by continous contributions to maintenance and operation activities either in labour, in cash, or in-kind. In some systems, the right of use can be forfeited, at least temporarily, when the individual users neglect their contribution and/or when users notoriously oppose the internal rules of water distribution exceeding the assigned stream of benefits. The individual shareholders can alienate their shares only by bequeathing them. Shares can only be leased or exchanged to other eligible members of the resource community. Unlike jointly owned private property „whose individual co-owners may sell their shares at will without consulting other co-owners", common-property regime shares are not tradeable or exchangable without the mutual consent of the shareholders.[91] This regulation makes it impossible that individual members can sell their shares, or may forfeit their shares upon change in residence, or may acquire shares through application but without purchase.[92] McKean assumes that the particular fashion of 'transferability' creates an inviolable bond between the co-owners and their rights, enhancing their interest in making the best possible use of such rights, and in operating with the longest imaginable time horizons. She argues further:

> It may be that the value of prohibiting the sale of rights to the commons has less to do with the effect on owners not allowed to sell their rights, and more to do with the undesirability of having any absentee owners contaminating collective decisions about the uses of common-property. Forbidding the sale of shares (...), especially to 'outsiders', is one way to guarantee that co-owners of the commons all have fairly similar economic objectives and will be able to reach agreement about how to use the commons.[93]

A variety of water distribution methods are found by which individuals can withdraw water from the system: fixed percentage (the flow of water is divided into fixed proportions); fixed time slots (each individual is assigned fixed time slots during which they can withdraw water); fixed orders (individuals take turns to get water).[94] They reflect the entitlement to a particular share of the common goods, and distribute products of the commons in direct proportion to private holdings of cultivated fields, i.e., they reproduce inequalities in private wealth. That the own-

---

[89] McKean 1992, p.258
[90] See Hussein et al. 1987
[91] McKean 1992, p.251; see the regulations on water rights in moslem law (Annex 12)
[92] Ibid., p.252
[93] Ibid., p.252
[94] See Yan Tang 1991, p.46

ers are co-equal does not mean that they are necessarily equal with respect to the quantities or other specifications. A higher share of the maintenance and provision costs are borne by those who benefit the most. As these are the wealthier members of the community, however, they can afford the greater costs. For the commons to survive, however, it is important that this pattern of distribution gives an incentive to the wealthiest co-owners to ensure that the common good is maintained and protected.[95]

Various strategies are set to deal with inequitable water distribution and the head-end/tail-end problem. The irrigation communities in the Philippines, for example, where each member has the right to farm a proportionate share of land, divided the area of one community into three or more large sections. By allocating each farmer a plot in each section all could farm some land in the advantageous 'head-end' location together with some at the 'tail-end', whilst keeping a fundamentally symmetrical position in relation to one another with rights to farm equal amounts of land.[96] The symmetrical position also enhances the members' interest to get the water to the tail-end plots. A more common strategy is that plots at the tail-end are assigned to the responsible officials as a compensation for their work. Again, the Philippinean Zanjeras provide „a positive reward for services rendered (and) also enhance the incentives for those in leadership positions to try to get water to the tail end of the system".[97] Wade reports that in Indian villages the common irrigators, and not the farmers, are responsible for water control up to the field level. The common irrigators are assigned to cultivate fields as a part of their salary which are located at the tail-end promoting their motivation to get the waters to the tail-end. A very common and widespread feature is that the canal patrollers/guards personally collect their share of grain from each of the beneficiaries, who can then withhold it if he perceives the provided services as poor.[98] Scattered holdings promote equitable water distribution, because an irrigator with land near the top end of one block may have another plot near the bottom end of another block, which diffuses the direction of the externality and helps to create a common interest in rules and organization. With unscatterd holdings, however, „the externalities of water use are 'uni-directional': the actions of irrigators with land at the head of the block impose costs on those towards the tail, but not vice-versa. In this case there is a clear interest between top-enders and tail-enders, the latter having stronger incentive than the former to agree to strong community organization and formal rules".[99]

The reviewed case studies describe consensual and cooperative risk-aversion strategies to ensure efficient resource use in times of water scarcity when demands exceed availability. There are increasing restrictions and limitations on assignments. In the Philippinean Zanjeras, where all individual irrigators have fields located in the upper as well as in the middle and the lower parts, the upper fields are irrigated first, then, subsequently, the middle and the lower parts receive water.

---

[95] McKean 1992, p.266
[96] See Ostrom 1990, p.83
[97] Ibid., p.83
[98] Wade 1988
[99] Ibid., p.185

Because all irrigators have fields in all parts of the irrigation area, „a decision about sharing the burden of scarcity can be made rapidly and equitably by simply deciding not to irrigate the bottom section of land".[100] Some systems stipulate which crops are to be irrigated. In Nothern Pakistan, for example, where water rights are linked to land, and vary with land use, in the list of priorities wheat and alfalfa, vital as winter fodder for livestock, come first, followed by fruit orchards, and lastly plantations of multipurpose trees interplanted with grasses. Priority is given to alfalfa, however, only if it is planted on cropland and not if it is interplanted with trees, as trees on cropland have 'junior' rights to water. When regular cropland is converted to orchards, however, it retains 'senior' rights. Even though cash returns from orchards are five times that from wheat, a mandatory eleven meter spacing from neighbouring fields acts as a constraint on horticulture development. Some farmers expressed resentment at what they consider to be anachronistic rules. Older farmers, however, who have experienced the need in the past to be self-sufficient, stressed the imperatives of food security, pointing out that trees also shade other crops, and that the long roots could also reach far into neighbouring fields to find vital moisture.[101] Case studies from Valencia (Spain) and Northern Pakistan show that the water needs of various crops are taken into account as another cooperative strategy when dealing with extreme water scarcities. Those crops which are in the most need of water are given precedence. There must be a common agreement amongst the group members about the relevance of crops, whether for the community's subsistence or market-oriented production. In Valencia the farmers are expected to consider other farmers' needs for water in times of drought. At the beginning of a drought period they should only apply water to those crops which need water most to shorten their turns for scarce water. The responsibility for determining how long a farmer may have water, according to the conditions of the farmers' crops and the need of others, is then increasingly taken over, as the drought continues, by the water master and his representatives.[102] In cases with symmetrical positioning of lands, or more or less equal farm conditions such as size and crops, compensations may not be necessary. Whether or not compensations are made to farmers receiving no water from those who receive water remains an open question.

Due to the pressure on individual users in times of water scarcity, the incentive for illegal water withdrawal is also present in farmer-owned and managed systems. Unlike public systems, the monitoring activities result in high rule conformance. The reasons are, as Ostrom et al. emphasize, that free-riders are easily noticed, and „these individuals' reputation as reliable members of the group, of considerable importance in small communities, are adversely affected. Thus, although overt sanctions are employed in user groups to reduce free-riding behaviour, the deduction results, in large part, from the increased information that all members have about each other's activities and the importance of a good personal relation in such settings".[103] Tang outlines that self-enforcing arrangements are adapted which

---

[100] Ostrom 1990, p.83
[101] See Dani and Siddiqi 1987, p.77
[102] Ibid., p.73
[103] Ostrom et al. 1993, p.136

require minimal supervision, and Bottrall stresses the relation between the higher acceptance of rules and sanctions, and the high quality of services.[104] The irrigators, themselves, monitor as a by-product. In Bontok villages in Northern Luzon (Philippines), a farmer or household representative usually directly observes the water distribution. There may be disputes between neighbouring paddy field owners concerning the division of water, but each farmer makes sure that a neighbouring field owner does not take more than his share.[105] Most of the systems employ specialized guards. Usually, the rewards to guards and monitoring irrigators in self-organized systems are that they are paid in-kind, more than in cash, according to the yield after the harvest is reaped; they collect their grain directly from each farmer, based on the amount of land which was irrigated; they have to justify themselves before others who are watching their performance in order to obtain their pay for the year; the pay may be withheld if the farmers are dissatisfied; to become re-assigned, their work has to be effective and impartial; they work in pairs (controlling each other); the area they serve does not connect to their own land. The rewards assigned to 'integrated' guards in self-organized systems are substantially and tightly connected to the work they do.[106]

All systems have rules for rule-breakers, and impose graduated sanctions. The execution of sanctions depends on the severity of the violation, i.e., if one notoriously breaks the rules, and on social considerations. In the traditional Bontok communities in Northern Luzon of the Philippines, disputes among villages occur primarily over the use of water, and there are numerous social mechanisms for promoting peaceful relations, e.g., the exchange of labour groups, and peace treaties between the villages. The distribution of water within a village is a particularly sensitive matter. It is considered the responsibility of every household to monitor its water supply closely to ensure it receives its fair share. Whenever a water theft is discovered, it is reported to other members of the irrigation system, who publicly scold the offender. The lack of stricter sanctions for such a serious offense may give the impression that the system is ripe for abuse, but feelings about irrigation water are so intense that more severe punishments, e.g., withholding water, are considered counterproductive, because they would divert precious labour from working in the fields to time-consuming litigations over water thefts. Moreover, it is believed that to withhold water would interfere with the basic right of every family to produce the food it needs for survival.[107] The Ifugao in Northern Luzon treat illegal water withdrawals as an offense punishable by fines. The first offense, when the culprit is discovered, is not punished; but there is a warning against repetition. If he repeats the offense, all the water is drained from his field, and/or he is beaten.[108] The Ifugao system of dispute settlement shows an interesting varia-

---

[104] Yan Tang 1991; Bottrall 1981
[105] Prill-Brett 1986, pp.54-84. Each Bontok village is an autonomous independent agricultural community with established claims in regard to land, water, hunting areas, and swidden gardens. The villages' population ranges from 600 to 3,000 persons.
[106] Weissing and Ostrom 1993, p.415; see also Ostrom 1992; McKean 1992, pp.272-275; Yan Tang 1991, p.47
[107] Prill-Brett 1986, pp.54-84
[108] see Barton 1969

tion with a particular incentive structure for the person entrusted with conflict resolution: it is a mediator system with so-called 'go-betweens'. No important transactions between persons of different families are completed without the intervention of this middleman. The go-between is the principal witness to a transaction. For this service he receives pay which is fixed, to a fair degree of exactness, for a particular service. This pay ranges from a piece of meat to a fee of twenty to twenty-five pesos. Go-betweens are responsible to both parties of a transaction, for the correct rendering of tenders, offers, and payments. Their word binds only themselves. The mediators are interested in a successful transaction in order to get their fee.[109]

The reviewed farmer-owned and -managed schemes show variations in their design of conflict-settlement institutions regarding the members involvement, the responsibility and range of discretion of the entrusted conflict mediators.[110] In most of the systems, the operational units have the responsibility of adjudicating conflicts, with binding decisions on the parties involved, while others have separate juridical bodies. Others have mediators, and binding decisions can only be made by a members' assembly. Whatever the design of institutions might be, they are accepted and legitimated by their members.

2.4
Maintenance Decisions in Irrigated Common-Property Regimes

Reviewed empirical studies show general patterns for maintenance activities in irrigated common-property regimes:[111]

The most important single factor is that only continous contributions to maintenance activities maintain the users' rights to irrigation water.

The maintenance obligations of the farmers relate to their individual user rights. In some systems, the burden increases with the distance from the main source. In Somalian and Pakistani schemes, work starts at the main canal and is carried out by all of the users. The individual users participate only so far, until they have reached their section. After that only those users continue whose fields are further away. Once the water reaches the tail, those at the head cannot be mobilized for repairs.[112] This rule acertains that the tail-end farmers, who have a vital interest to get the water down, contribute to a high water conveyance efficiency, and they will then exercise strict control on the head-end farmers to get their fair share. Ostrom assumes that whether equal contributions are required from the members, or whether the rules define maintenance contributions in proportion to benefits depends on maintenance intensities.[113]

---

[109] Ibid.

[110] For other conflict-resolution arrangements see Ostrom 1990

[111] See Boguslawski 1986; Seger 1986; Dani and Siddiqi 1987; Hussein et al. 1987; Abdellaoui 1987; Wade 1988; Cruz 1989; Ostrom 1990; Yan Tang 1991; Hilton 1992; Ostrom et al. 1993

[112] For the regulations in the Ottoman water law, see Annex 12

[113] Ostrom 1993

The decisions on maintenance are made by the legitimate decision-making body of that resource community which has the authority to designate the irrigators who need to participate in maintenance works, and which has the right to impose fines on those who refuse to do so.

Local resources are mobilized for maintenance, and they fully recover recurrent costs. The beneficiaries are willing to contribute a high amount of labour to sustain their systems.

The contributions may be carried out by the members themselves or by hired, and paid, labourers. Minor repair work may be done by guards, but any significant breach may result in the mobilization of the entire user group.

In some systems, the control on maintenance contributions are somewhat higher than the control on water distribution, and non-contribution is fined.

The decisions on the timing of maintenance activities are made by the farmers, or their representatives, and thus conflicts with farm operation plans, and correlated opportunity costs, can be reduced. Gathering information on local maintenance requirements for certain channels is done by the canal patrollers and the farmers themselves. Because each farmer is interested in getting the water to his plot, it can be assumed that breaks, leakages, flooding, etc., are immediately reported.

Finally, social mechanisms exist to lower the burden on poorer households, 'helpless' families, or households without males. In such cases, all village households share the extra work equally.

3.
Improving the Performance in Large-Scale Public Irrigation Systems

The fiscal crisis in the developing countries has prompted the governments to initiate irrigation management transfer programmes. In some countries, for example, Mexico, Columbia, and the Philippines, state-owned and -managed systems were transferred to different kinds of water user associations. In Columbia, the management of two irrigation districts was turned over. The Philippines had a three-stage transfer programme of which the last stage included the responsibility for all but the main storage, diversion and conveyance works. In Mexico, 77 irrigation districts, each with a size of about 3,000 ha, were transferred to user groups.[114] But complex large-scale public irrigation schemes cannot, however, always be replaced by farmer-managed systems, and water-user organizations were stimulated within the service area of government-managed irrigation systems where user organizations and state agencies cooperate in management issues. Recent discussions concentrate, foremost, on changing finance policies in relation to institutional structures, because it is believed that the effects of cost recovery, i.e., improved operation and maintenance, depend on whether an irrigation agency is

---

[114] The World Bank 1993, p.105 (see also Appendix C. Privatization and User Participation in Water Resources Management); see also Turral 1995, p.2

financially autonomous, (or semi-autonomous), or financially dependent on resource allocations through the governmental budgetary process.[115] The relationship
between the irrigation agencies and the higher-level goverment authorities, and the
relationship between the direct beneficiaries and the irrigation agencies are subject
to changes. High emphasis is given to the participation of farmer groups in large-
scale public irrigation systems, where water-user groups could/should play a significant role in resolving water allocation and maintenance problems. There is a
wide range of options for jointly managed public-farmer irrigation systems.[116]
Subramanian et al. differenciate several forms of joint management which are
based on which entity has responsibility for regulation, ownership, operation and
maintenance, and user representation.[117] The main differences concern the state
agency, e.g., whether it is financially dependent or autonomous; the design of the
participating farmer groups, e.g., their internal structure, their range of decision-
making, the incentives for participation, and how interaction and cooperation
among the state agency and the user groups is institutionalized.

3.1
Financially Autonomous or Semi-Autonomous Irrigation Agencies

There is evidence that operation and maintenance improve with financially
autonomous management units who rely on the water charges collected from the
beneficiaries. An early evaluation report of the World Bank states that „in general,
the best irrigation performance was achieved in projects where a) the irrigation
agencies themselves were responsible for the collection of the financial charges,
and b) the fund collected by the irrigation agencies remained with them for the use
in operating and maintaining the irrigations projects".[118] The agencies should have
control over the resources collected from the beneficiaries and be able to decide on
the allocation of resources for O&M. The beneficiaries then control the demands
for O&M, not the government, thus introducing accountability. The agency depends on the farmers' 'willingness to pay' for good services, and the beneficiaries
may use the threat of non-payment as a means of leverage over the agency. On the
other hand, cost recovery places financial responsibilities on the beneficiaries, and
lowers the burden on the central treasury. The National Irrigation Administration
(NIA) of the Philippines provides a famous example of a financially autonomous
irrigation agency. The NIA is a government corporation with a board of directors
that includes the Administrator of NIA, the Minister of Public Works, the Minister
of Agriculture, the Minister of Economic Planning, and the General Manager of
the National Power Corporation. As a government corporation, it has the authority

---

[115] See Small et al. 1989; Feder and LeMoigne 1993; Frederiksen 1992; The World Bank
1993; Repetto 1986; Bottrall 1981; Livingston 1992
[116] The term 'public-farmer systems' is used here for schemes where management activities are jointly undertaken by farmers and irrigation staff with official recognition.
[117] The forms are (a) agency O&M, user input; (b) shared management; (c) Water User
Organizations O&M; (d) Water User Organization ownership, agency regulation. (for
details, see Subramanian, Jagannathan and Meinzen-Dick 1995, pp.50-51)
[118] Duane 1986, quoted in Small 1989, p.15

to collect water charges from the beneficiaries for the irrigation services it provides. The NIA is allowed to impose charges to generate sufficient revenues to
cover O&M costs and to recover initial investments (within 50 years). The government subsidizes the cost of interest. Of critical concern to the NIA is the balance between the revenues it receives in the form of irrigation service fees that are
collected, and the expenditures it incurs for O&M. The irrigation service fees
collected in 1979 and 1980 were equivalent to nearly 70% of the O&M costs, but
dropped to only 50% in 1982 when O&M expenditures increased sharply, and
collections declined. Factors that affect the rate of collection are adequacy of personnel and budget; communication among personnel and farmers; the capacity of
the irrigation system to perform adequately, and performance evaluation. Finance
policies have put considerable pressure on the NIA to reduce the deficit which it
encounters in its operation either by increasing revenues or decreasing expenditures.[119] Given that the NIA has not followed the undesirable strategy of reducing
expenditures by drastically curtailing services and letting irrigation systems deteri-
oriate, most of the options open to the NIA involve placing greater responsibility
on the farmers,[120] and emphasizing better maintenance and de-emphasizing the
construction of new projects.[121]

Another example are the financially semi-autonomous parastatal irrigation associations in Taiwan. The managers of the associations are encouraged to manage
the systems efficiently, because only a high rate of fee collection can preserve the
jobs for the staff. The irrigators can put considerable pressure on the managers by
withholding their payments when they perceive the service as poor and the water
receipts unreliable. In some years, the collection efficiency ranged to about 98% of
the assessments, whereas in other years it has fallen to about 28% „in part due to
the inability of farmers to pay the charges, and in part because of farmers' unwillingness to pay".[122] Similar experiences have been reported from China and India.[123]
It is assumed that the decisions on operation and maintenance made by public
officials in O&M divisions, and particularly in local units, are more likely connected with the interests of the potential beneficiaries, and that the incentives facing public officials depend strongly on the institutional structure within which the
O&M units are embedded. If the units responsible for maintenance activities are
not a part of a large civil engineering hierarchy oriented to construction, and if the
fees paid by the farmers account, to a considerable extent, for the financing of
maintenance, then their decisions are linked to the local farmers as it is in Taiwan,
South Korea, and the Philippines.[124] In Taiwan, O&M is separately organized from
engineering, and is the responsibility of parastatal organizations. The fees paid by
the farmers provide the major share of their budget, and the communication is
closer and more effectively linked to the interests of the local farmers (an irrigation

[119] See Svendson 1993
[120] Small and Adriano 1989, pp.189-228
[121] Cabanilla 1984
[122] Bottrall (1978, p.65) quoted in Small 1989, p.15
[123] Small et al. 1989; Repetto 1986
[124] Bromley et al. 1980; Wade 1987; Weissing and Ostrom 1993

group chief is elected, usually a local farmer, who supervises water distribution, conflict management and maintenance).[125]

With autonomous, or semi-autonomous agencies, the constraints set on the management, like rigid rules and controls that inhibit staff initiatives in adapting cost-effective managemant solutions, can be offset. On the other hand, in cases where it is common that the governments pay for the capital costs while the users bear the current costs, an incentive is created to the farmers „to accept deferment of certain types of maintenance to the point that the work can be considered to be rehabilitation of the system".[126]

For financially dependant irrigation agencies, Bottrall et al. suggest introducing some degree of competitiveness among irrigation projects; indicators for performance quality could be introduced against which financial increments could be earned, e.g., levels of production, equity of water distribution, levels of recovery of water charges. The irrigation agency could be provided with a fixed grant and a variable bonus. Or, alternatively, „each project may be allowed to retain a certain percentage of the revenue it had raised in accordance with its performance, to which the flat-rate grant could be added".[127] Experiences in Korea show that in financially dependent agencies, performance indicators and agreements can be set to permit comparative evaluation of the short- and long-term performance of public officials. Performance agreements are being introduced, which should ensure that information is available for the evaluation, that rewards to managers and employees are linked to their performance, and that the evaluation is made by independent auditors. To increase accountability to the users, the performance-based ranking of public companies is published in the press. The best manager not only gets prestige, but monetary compensation.[128] If the ministries' O&M wings remain dependent on central governments, however, problems of defining performance indicators and performance measurement, monitoring and control remain. Other options are discussed for improving the quality of management of which one is breaking the defacto monopoly of the state by involving the private sector, for example, in maintenance works. In the Gezira scheme in Sudan, silt removal was contracted out to a parastatal organization, and improvements are thought to be possible with private sector participation.[129] Above all, the political will on the side of the government for improving performance also remains a precondition.

---

[125] See Moore 1988, in: Weissing and Ostrom 1993, p.418
[126] Small and Carruthers 1991, p.72
[127] Bottrall et al. 1982, pp.53-54
[128] For details, see World Bank Development Report 1994, p.43
[129] See Plusquellec 1990

3.2
Participation of Water User Organizations

With a deficit of financial resources and poor performance, the farmers' contributions to operation and maintenance is found to be the most appropriate solution.[130] Evaluation reports have conceded that, under certain conditions, „water delivery services improve because farmers have stronger incentives to distribute the water and better information about irrigation needs. This permits more flexible allocation patterns and more careful monitoring of actual deliveries. System maintenance improves when (water user organizations) have a greater stake in the systems. Farmer members are more likely to monitor the condition of irrigation structures and less likely to damage them if the (water user organizations) must bear the costs of repair".[131] In Tang's study, for example, among the bureaucratic cases, those with local farmer organizations tend to have higher degrees of rule conformity and improved maintenance of the delivery network.[132] Wade reports on jointly managed systems in South Korea that „the standard of maintenance seems, impressionistically, to be good in the farmers' delivery channels, and in many places less good in the main and branch channels", while the drainage channels are maintained with less care and diligence. Maintenance rules and control on free-riders are reminiscent of those of self-governing systems. The customary 'spring-cleaning' of the distribution channels from silt and refuse is carried out on the same day by all farmers, with each farmer cleaning the part of the channel that borders his land rather than operate collectively. This work only takes each farmer a few hours because of the small size of the holdings, but should he fail to carry out the work within a reasonable time, and thus impair the conveyance of the channel, various informal pressures will be put on him. Many will frequently and directly ask him to do so, and the matter could be brought up to the subvillage or village meeting.[133] The physical and financial responsibility for the tertiary-level facilities in Indonesia are subject to the farmers through their local institutions, such as the village government and various types of water user associations. These associations usually require that farmers pay a fee per hectare, per season, either in cash or in-kind. In addition, farmers also contribute materials for construction and labour, as the need arises. Small and Adriano assume that „the associations are successful in collecting the membership fees (...) because they are able to implement the regulations and impose sanctions agreed upon by the farmer-members".[134] For Indonesia as well as for South Korea, the authors agree that where the tertiary level is within the responsibility of the users, the performance has improved, unlike that of the main system operation where the funds provided are still inadequate, and the performance is still poor.

---

[130] The World Development Report 1994 (Infrastructure for Development) strongly emphasizes that "participation increases water project effectiveness by improving maintenance". (1994, p.76)
[131] Subramanian, Jagannathan and Meinzen-Dick 1995, pp.11-12
[132] See Yan Tang 1991, p.48
[133] See Wade 1982, pp.70-72
[134] Small and Adriano 1989, p.54

Farmers' participation in operation and maintenance is desired by some governments because of its positive effects on O&M expenditures. The farmers, or their organizations, are expected to do the work that needs to be done on the systems. Reviewing farmers' involvement, Carruthers and Morrison wonder „to which extent this is wishful thinking of those remote from the field level problems, or a reaction to shortages of government revenue, or simply a switch of tactics from centralised planning approaches to devolved participative mechanisms in line with a widespread current political fashion sometimes summarized as 'subsidiarity'".[135] The farmers themselves may have no interests in operating and maintaining. Carruthers attributes the farmers' bias to the issue of location on the watercourse, because „those at the head of watercourses have very little incentive to support maintenance efforts below them since they have ample water. For those at the tail, there is very little incentive because even if maintenance is performed they have no leverage over those at the head".[136]

It remains an important issue under what conditions farmer participation can achieve success, and how the organizational setting should be designed in a public-farmer management structure. It is, however, agreed that there must be incentives for the farmers to participate, and there must be benefits for governments and state agencies to implement transfer programmes and support participation. The latter are easier forthcoming, at least, the local recovery of O&M costs lower the burden on central treasuries. Chambers sets some conditions which, in his view, are necessary for collective action to be forthcoming: They are that „the group must have a common hydrological interest; water must have a high value to the group; the group's water supply must be unsatisfactory (inadequate, inconvenient, unpredictable, untimely, etc.); farmers must perceive a reasonable chance to improve it".[137] Uphoff et al. concede that benefits for the irrigators from participation „are likely to be greatest over the middle range with regard to water availability. With surplus available, it is not worth the time and effort of organising user groups since all farmers are receiving adequate water, and with scarcity there is often too much conflict to allow users to form agreements".[138] Other scholars assume that the community must gain more than just the costs involved; services could be provided more cheaply; participation could ensure that things are done in the right way, and indigenous knowledge could be used.[139] Bottrall defines principal conditions for small group formation and duration in large-scale public systems, some of which are identical to Ostrom's working rules (see Box 2.3.).

There is no agreement in the scholarly discussion on participation. Participation is regarded as „the inclusion of the intended beneficiaries in solving their own problems",[140] and Yan Tang and Ostrom emphasize that „participation in management is not enough. Farmers have an important role to play in the governance of the systems. Governance is defined as the establishment of specific working rules

---

[135] Carruthers and Morrison 1994, p.49
[136] Ibid., p.52
[137] Chambers 1988, p.171
[138] Quoted in Carruthers and Morrison 1994, p.50
[139] See, for example, Smout 1990
[140] McPherson and McGarray 1990, p.87

used to allocate water, assign responsibilites for labour and monetary resource mobilization, to resolve conflicts, pay workers, and sanction non-conformance" within a clearly defined area of responsibility.[141]

Box 2.3. Small group formation in large-scale public systems[142]

"The *first* condition for success is that it must be possible through group action to secure a 'collective good', i.e., one which can benefit each individual member but which is realisable only if he collaborates with others in an organization to obtain it.
Secondly, the net private benefit to which each person is given access through membership of the group (taking into account the transaction costs and loss of individual discretion which membership involves) must exceed what he can obtain by other means.
Thirdly, the nature of the group's activity should preferably be such that individuals, in pursuit of their own private benefits, are inclined to do so in a manner which benefits the group as a whole and promote its long-term growth.
Fourthly, each member of the group must agree to share the benefits it yields on the same terms as his fellows: the basis of group membership must be a reciprocal one.
Fifthly, there must be agreed sanctions, which may ultimately have to be supported by government, to protect the group against external harassment or private exploitation by any of its own members.
And *finally,* the group must be of a size which is appropriate to its functions and the management capacity of its members. (...) interest is generally much easier to secure in relatively small groups."

The latter condition is of major importance for successful group action, because continous state interventions might be one reason for disruptions and/or lack of action.[143] Clearly designated responsibilities exist for the maintenance of the main and branch channels in South Korea; the labour is divided, with the staff of the Maintenance Section acting in a supervisory capacity, and with patrollers and the farmers in a labouring capacity. Spontaneous actions emerged on the lower distribution and drainage networks; although the state agency conceded that the lower levels are within the responsibility of the farmers, Wade reports that farmers say the state agency is meant to be responsible for the maintenance of all channels, but the farmers have to do it because of the agency's default.[144] In order to avoid those black spots where nobody feels, or has been made, responsible, and where counter-incentives and conflicting targets impede successful actions, Huppert and Urban developed an analytical instrument, the Service Interaction Analysis, to identify the participants, services and interactions between service producers and their clients in a multi-organizational network.[145]

Whether the performance effectively improves with the participation of farmers' groups is a matter of how the relationship between the state agency and the

---

[141] Yan Tang and Ostrom 1993, in: Morrison 1993, p.56
[142] Source: Bottrall 1981, p.216
[143] See IIMI 1987
[144] Wade 1982, pp.70-72
[145] Huppert and Urban 1994

farmers' groups, and within the farmers' groups themselves are designed. A matter of particular importance is what incentives are provided for the participation of the farmers to contribute towards operation and maintenance activities. Subramanian et al. reviewed internal and external conditions for sustainable water user organizations.[146] Key features for their organizational structure are origin, membership definition, size, federation and leadership, and specialization; physical and technical factors, social and economic factors, and policy and governance factors exercise important influence on the conditions under which water user organisations are likely to succeed.[147] Some scholars assume that reduced negative environmental externalities similarly result from improved irrigation services under the management of water user organization, but empirical evidence is not proven. No research systematically issues the effects of farmer participation in jointly managed systems on groundwater and soil conditions.

4.
Potentials and Constraints for Controlling High Groundwater Levels and Salinization

The differences between large-scale public and small-scale farmer-managed systems are substantial (see Table 2.1.): The reviewed literature has shown that governments and their bureaucracies have not been able to successfully enforce the rules for water allocation and maintenance, and for their financing. The causes for poor performance refer to budgetary as well as to non-budgetary constraints. As some means for controlling high groundwater levels and salinization such as the regulation of water inputs by the water allocation rules and investments in delivery systems with low percolation coefficiencies, and their maintenance, coincide with means to achieve efficient water allocation, others do not. The most important single factor are investments in adequate drainage systems and their sustained maintenance; others are the rationing of water by means of economic incentives and the internalization of negative externalities generated by salinization. Experiences in many countries have, however, shown that state regulation has failed.[148] Where groundwater and salinization control means coincide with means for improving water allocation, the previously mentioned constraints, of course, also apply to them. Public investments in adequate drainage systems and their maintenance, however, is a much greater problem which needs to be explained. In these particular fields, research is poor, and groundwater and salinization control in large-scale public irrigation systems has not been evaluated as a task on its own.

Irrigated common-property regimes show high performance levels in water supply and the maintenance of delivery channels, which are important groundwater and salinity control means. Low-cost investments prevail, and investment decisions

---

[146] For details, see Subramanian, Jagannathan and Meinzen-Dick 1995, pp.13-49
[147] See Subramanian, Jagannathan and Meinzen-Dick 1995, pp.13-49
[148] Indicators for this failure are the increasing rehabilitation needs of many public irrigation systems which include the construction of new drainage systems and the modernization and repair of older ones.

are within the labour capacity of small farms. Local resources are mobilized, and recurrent costs can thus be fully recovered with high labour inputs for maintenance, compensating for the low-cost investments. The regimes can create institutional certainty through strict rule enforcement. They effectively limit free-rider behaviour, and the potential for shirking and rent-seeking by officials is low. The rules for water allocation reflect the claims of the individuals, and their enforcement guarantees an equitable supply. Built-in mechanisms also ensure that water reaches the tail-end plots. Water supply decisions are able to respond to local conditions and needs as information costs are low. Whether the supply is reliable and adequate is questionable, but it is predictable, because the individuals are informed about water availability. High water inputs as an individualistic strategy to overcome uncertainty is restricted by rule enforcement. Whether water inputs in times of water abundance are designed to limit the percolation of water has not been investigated. Restrictions on water withdrawal are introduced with increasing water scarcity. Then, additional cooperative and consensual risk-aversion strategies come into force, securing efficient water use, e.g., particular crops are given precedence, and priorities for land to be irrigated are defined.

Although operation activities and the maintenance of the delivery systems may improve with the participation of farmer groups, in jointly managed public-farmer irrigation schemes, drainage systems are still poorly maintained by the state agency and the farmers, and means of water-saving are not put to question. Whether demand on water savings and drainage is transformed into policies depends on political decisions, and on the incentives to achieve the state agencies and farmer groups interest in maintaining the drainage networks. It is presumed here that if available water could generate higher returns, whether within the agricultural or in another sector, i.e., energy, municipal water supply, industries, then water-saving policies would be more likely to be initiated by politics. If distribution problems occur within a project area, agricultural investments may be underutilized, having impacts on exports and food production, and probably necessitating the increase of imports, and the possible lowering of revenues from income. High groundwater levels and salinity could generate costs in the form of lower yield and incomes, with negative impacts on the national economy as well as on farm economy. Though, the central questions for the analysis of the case study are: who expresses demands on water savings and drainage? do the negatively affected parties have the means to set the issues on the agenda? how is collective action forthcoming? and who has to bear the costs deriving from high groundwater levels and salinization? or to put it differently, what are the institutional conditions for the effective control of high groundwater levels and salinization?

Table 2.1. Differences between large-scale public and small-scale farmer-owned and managed schemes with regard to investment, and in operation and maintenance

	Public Schemes large-scale and high number of farmers	Farmer Schemes small-scale or low number of farmers
Investment in storage and distributory networks		
Investment costs	High Central budget and external funds	Low Local resources within labour capacity of small farms, or hired personnel
Decision-maker	Government, ministries	Farmers
Construction	State agency	Farmers
Repayment	Highly or totally subsidized	Recovery through local resources
Investment in drainage networks		
Investment costs	High Central budget and external funds	n.a.[a]
Decision-maker	Government, ministries	
Construction	State agency	
Repayment	Highly or totally subsidized	
Maintenance of delivery networks		
Decision-maker	Government, ministries, state agencies	Eligible members or their representatives
Financing	Central budget allocation; costs per hectare and crop charged to beneficiaries High rate of free-riders, low repayment to central treasury	Local resources; individual inputs are related to user rights/shares; socially balanced Low rate of free-riders, total recovery of costs
Programme Implementation	Annually routinized programmes Local O&M branch	Local requirements Eligible members, guards, or hired labourers
Performance	Low water conveyance efficiency, but better than maintenance of drains; low labour productivity	Good, or sufficient
Control	Weak	Strict

Table 2.1. (continued)

Maintenance of drainage networks

Decision-maker	Government, ministries, state agencies	n.a.[a]
Financing	Central budget allocations; costs per hectare charged to beneficiaries; High rate of free-riders, low repayments to central treasury	
Programme Implementation	Annually routinized programmes Local O&M branch	
Performance Control	Worse than delivery systems Weak	

Operation

User rights to water	All farmers in a service area → landownership or lease; not transferable	Eligible members → initial contribution, inherited, leased or bought, performing continously their duties Adhered to land or separated; sometimes transferable
Water allocation rules	Imposed	Consensually decided by members
Abundancy - scarcity	Unspecified water reductions	Proportional water reductions and/or cooperative risk-aversion strategies
Operational body	O&M division: public officials, hired labourers, ditch riders	Elected or appointed local officials or inherited position; with or without guards
Financing	Central budget allocations; costs per hectare and crop charged to beneficiaries High rate of free-riders, low repayments to central treasury	Cost recovery through local resources from beneficiaries; Low rate of free-riders, high repayments
Performance	Poor (inadequate, unreliable, unpredictable, inequitable)	Good (high predictability, equitable)
Rule enforcement	Weak (low monitoring, no enforced sanctions)	Strong (high monitoring, enforced sanctions)
Control of performance	Weak: higher level authorities and within agency Potential for shirking, rent-seeking, corruption	Strict: through members; guards Low potential for shirking, rent-seeking, corruption
Conflict settlement	Public officials; local court	Juridical body; members' assembly; operational unit

[a]   not applicable

Chapter Three
The Implementation Process of the Lower Seyhan Irrigation Project, with Special Reference to Means of High Groundwater and Salinity Control

In Turkey, in the 1950s, cultivation could not be extended because most of the land suitable was already developed; on the contrary, cultivated lands had to be limited to 16.5 million ha instead of 24 millions due to striking erosion problems. Land with high potentials had to be used more intensively to secure food production and to increase export potentials. Of approximately 6 million ha of land that was estimated to be economically irrigable only 1.2 million ha, or 20%, were irrigated, with 330,000 ha irrigated through state-constructed facilities, and 873,624 ha constructed and operated by individual farms, local organizations and private enterprises.[1] With the adaption of a new constitution in 1961, the Turkish government continued to favour a state-induced economic and social development strategy directed towards overcoming regional imbalances. The State Planning Organization (SPO), established in 1963, coordinated economic sector development by defining priority areas and sectors, and deciding on the provision of financial resources for the public sector. The First Five Year Development Plan was prepared for the period 1963 to 1967, to promote a development rate of 7 %. Within the agricultural sector TL 11.3 billion was invested, of which 45.3 % was allocated to irrigation development activities, bringing more than 500,000 ha under irrigation, and rehabilitating existing projects.[2] The Çukurova region was selected for development with the Lower Seyhan Irrigation Project as its centrepiece. It was, with its 175,000 ha, the largest irrigation project in Turkey.

In the 1960s, prior to the installation of the irrigation project, rainfed agriculture was practised on 95% of the project lands, with cotton accounting for about 85% of the cropped area, and wheat for most of the rest. The yields were low due to the continous monocropping of cotton, inadequate moisture during the growing season, and insufficient fertilizer use. The main objectives of the irrigation project were to increase agricultural production on 175,000 ha of fertile land, and the development of the full potential of the region. The climate conditions favour cropping throughout the year and the planting of a wide diversity of crops. The project would improve the yields of cotton and wheat, as well as provide the benefits expected from the diversification of agricultural activities with high value

---

[1] See Uner 1967
[2] See Onat et al. 1967

fruits and vegetables. Diversifying the cropping pattern was also thought of as a means to achieve sustainable high yields, while monocropping would only increase insect infestations and other disease problems. The cropping intensity could achieve about 140% with double cropping. It was expected that about 45% of the project area would be used for producing non-irrigated winter crops, such as dry beans, peas, and winter vegetables like melons, watermelons, and cabbage. The net incomes per hectare of the farms were expected, at full development, to increase 8 times, from TL 410 to TL 3,330. Accordingly, the use of modern inputs, especially high-yield varieties of seeds, fertilizers and agrochemicals would be introduced. Although it centred on agricultural development, the project would be a motor of regional growth for agricultural and other industries and commerce.[3] The city of Adana was already an important industrial centre with continous immigration from outside the area, and the annual population growth rate was above the Turkish average. Temporary movements of labourers into the area always reached a maximum during the cotton-picking season.

This chapter presents the natural conditions in the Lower Seyhan Plain,[4] which tend towards high groundwater levels, and then the implementation process of the Lower Seyhan Irrigation Project whose hydrological infrastructure facilities and on-farm development works were installed in phases. The analysis concentrates on the stage I and II projects which had been financed with external credits from the International Bank for Reconstruction and Development (IBRD) and the International Development Association (IDA). The stage III project was financed solely from internal resources, and was part of the annual investment programmes over a period of ten years. Finally, the rehabilitation projects and the planning documents for the yet to be commenced stage IV project area, and the impact of the subprojects on groundwater and soil conditions are documented. The early planning documents, the preliminary studies and the evaluation reports are reviewed giving particular regard to the control of high groundwater levels and salinity, and how initial investments and the implementation process addresses on its alleviation.[5]

1.
Natural Conditions in the Lower Seyhan Plain

The Lower Seyhan Plain, the area of the Lower Seyhan Irrigation Project, is bordered in the north by the Taurus Mountains, in the east by the River Ceyhan, in the west by the River Berdan, and in the south by the Mediterranean Sea. The plain is divided by the River Seyhan into the Yüregir and the Tarsus Plain. (see Map 3.1. The Çukurova Region)

---

[3] See DSI and TOPRAKSU 1961; IBRD/IDA 1963
[4] The geographical term 'Lower Seyhan Plain' is used for a part of the Çukurova region laying south of the city of Adana, bordered by the two rivers Berdan and Ceyhan in the west and east. (see Map 3.1.)
[5] The term 'implementation' is used for the process of executing investments, and includes the planning and construction activities, and the finance modalities.

Map 3.1. The Çukurova Region[6]

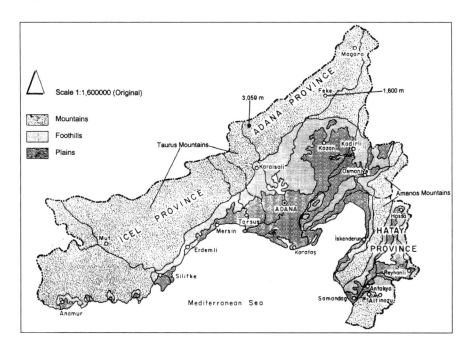

The region is one of the most fertile areas of Turkey. The climate of the Lower Seyhan Plain is mediterranean, with very hot and dry summers, mild and wet winters. The average annual rainfall is 640 mm; the distribution of rainfall over the year is 21% in autumn, 45% in winter, 25% in spring and 5% in summer. The humidity is relatively high throughout the year. The annual average evaporation is 1557.9 mm (1963-1984), of which 64% evaporates between May and October.[7] (see Fig. 3.1., 3.2. and 3.3.) The average temperature is 18.8°C (1929-1985). Frosts occur, statistically, once in three years, lasting not more than 3 to 4 days, and strong frosts occur in every 20 to 30 years. The frequency of temperatures of minus 5°C is statistically once every 4.5 years. Soil temperatures below 0°C were not yet registered.

---

[6] Source: Consortium TAHAL-ECI 1966, E (modified)
[7] Adana ve Karatas meteoroloji istasyonlarinda yagir, sicaklik ve buharlasma

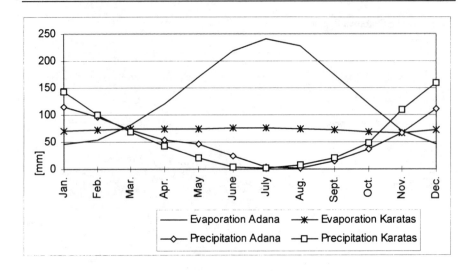

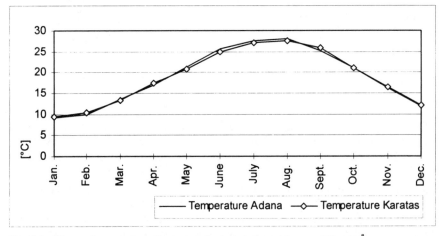

Fig. 3.1. Climate diagrams from the meteorological stations Adana and Karatas[8]

---

8   Ibid.

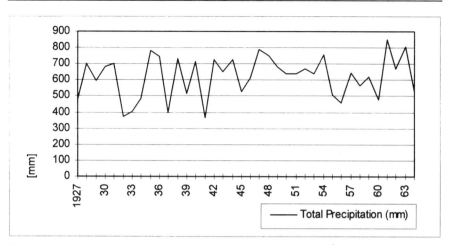

Fig. 3.2. Mean annual precipitation at Adana station in mm (1927-1964)[9]

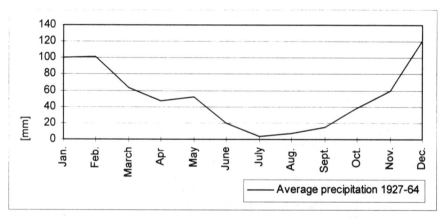

Fig. 3.3. Average monthly precipitation at Adana station in mm (1927-1964)[10]

The climate is favourable for cropping throughout the year, and accomodates many crop species, i.e., cotton, citrus, cereals, vegetables, etc.. The Taurus Mountains, which border the delta in the north, create a natural barrier for the northern winds. The temperatures are therefore usually higher than at other locations on the same latitude. The topography of the delta is very flat, with a maximum elevation of 60 meters above mean sea level in the north, with the minimum ranging from zero to 0.4 meters in the south. The general slope stretches to the southwest, with grades varying from 1% in the north to 0.1% or less in the southern parts bordering onto the Mediterranean Sea. The Taurus Mountains, bordering the Çukurova in the

---

[9] Ibid.
[10] Ibid.

north, are of Cretaceous limestone which show superficial fractures due to weath-
ering. The southern foothills of the Taurus consist mainly of limestone, and of
marine deposits of the Miosene age which form sandstone, marls and shale, over-
laid by Pliocene conglomerates and sandstone. In the south of the foothills alluvial
deposits from the Quaterny age are found. The delta, south of the European Highway 5 and of the city of Adana, consists of recent deposits of the three rivers Seyhan, Ceyhan, Berdan and their tributaries. These deposits are composed of clay,
silt, sand and gravel, with the clay components dominating. The plain is sur-
rounded by Miocene and Cretaceous hills to the southeast and east. Continous
sand dunes are found along the eastern seashore of the Yüregir Plain, and swamps
and lagoons have been preserved along the northern border of the dunes (the Salt
Sea, and the Akyatan and Agyatan Lagoons). The soils of the higher elevated lands
and of the hills are residual, red-coloured, and heavy-textured Terrae Rossae.[11]
They are mainly shallow, with a depth ranging from 30 cm to 120 cm, and are
composed of conglomerate and limestone formations. In the main areas of the
central and lower delta alluvial soils have developed on the deposits of the three
main rivers and their tributaries. These are the deepest and most fertile soils, with
depths varying up to 20 m. Solonchaks, Solonetzes, Grumosols and Chernozems
also can be found in the delta regions.[12]

The River Seyhan, with its tributaries, supplies the project discussed here with
irrigation water. The river's watershed comprises about 19,300 square kilometers.
The planning of water resource development on the River Seyhan started in 1939.
In the 1940s, a diversion dam, flood control barriers, and two main conveyance
channels were constructed from which 18,500 ha could be irrigated. In 1956, the
Seyhan Dam was completed and the hydroelectric power plant started its operation
with an installed capacity of 54 megawatts, and an average annual power generation of 350 Gwh. The Seyhan Dam and its reservoir also serve as a flood control
for 24,500 ha of agricultural land and the city of Adana, and the area surrounding
the reservoir is used for recreation. The river's average annual accumulated discharge is between 5.5 to 6.0 billion cubic meters.[13] The available volume of surface water from the River Seyhan, dammed in the Seyhan reservoir, is sufficient to
irrigate the project area of 175,000 ha,[14] and the water quality is most suitable for
irrigation purposes. The average electrical conductivity (EC_i), i.e., the value which
indicates the salt content of the irrigation water, is 388.9 micromhos/cm with a pH-

---

[11] See Topraksu 1974; Dinc et al. 1990
[12] See Ibid.
[13] DSI, Genel Müdürlügü 1989
[14] The storage reservoir has a maximum active storage capacity of 745 million cubic
meters. Between 1956 and 1986 the active storage capacity was reduced by 60 million cubic meters, which affected the flood control capacity, but not the capacity for
irrigation and energy generation. It is estimated that approximately 12 million tons of
sediment discharge annually into the reservoir, partly accumulating in the lake. The
upstream Catalan Dam was therefore constructed with a high dead volume capacity,
and, in the 1980s, the Cakit-Cayir Project implemented erosion control means in the
upper catchment area of the Seyhan and its tributaries.

value of 8.0.[15] The electrical conductivity of the water in the Seyhan reservoir ranges from 180 to 525 micromhos/cm; the sodium adsorption ratio (SAR) is low, ranging from 0.3 to 1.5.[16] 87% of the samples from the river were classified as C2-S1; the others were classified C1-S1. This indicates that there is a very low sodium hazard, and a low to medium salinity hazard, which can be disregarded as the winter rains readily counteract the slight tendency of salts to accumulate during the summer, provided, of course, that there is adequate drainage.[17]

The annual accumulated groundwater volume in the deeper groundwater storeys, which reach a depth of about 300 meters, is estimated to be 30 million cubic meters.[18] Artesian groundwater is not used for irrigation purposes, because this is not economically feasible, but is explored for drinking water. The intake area for this water is in the foothills to the north of the plain, and the aquifers are mostly confined, i.e., under artesian pressures. The confining layer is generally 20 to 100 meters thick, slightly permeable, and allows the slow upward seepage of the water. An impermeable soil layer separates these storeys from the accumulated groundwater near the surface.[19] As a result, a superficial water horizon has been established near the ground surface. The water level is high, reaching a maximum level of 0.5 to 1.0 meter below the land surface and a minimum of 2 to 4 meters. Large areas become waterlogged during the winter, or have a water level which is less than one meter deep. This level approaches or exceeds the ground surface in winter, and drops during the summer because of evaporation, depositing detrimental salts near the ground surface. Groundwater near the soil surface is recharged by surface and subsurface sources, e.g., precipitation, intrusion from the rivers, inflows from outside the plain, and infiltration from sea water near to the coast. By far the most important recharge source is precipitation.[20] Precipitation is highest between December and February, and the recharged groundwater reaches harmful levels in many areas. Surface pondage in numerous depressions and flat areas

---

[15] Kanber and Yüksek 1979, p.4 (Tab.5); see Elektrik Isleri Etüd Idaresi Genel Müdür-lügü 1989, pp.95-98

[16] The electrical conductivity (EC) of water is the reciprocal of electrical resistance measured at a temperature of 25°C (micromhos/cm). EC-values are differenciated for irrigation water (EC_i), groundwater (EC_{gw}), drainage water (EC_d) and for the soil saturation extract (Ec_e). The Sodium-Adsorption Rate (SAR) is a ratio for soil extracts and irrigation waters used to express the relative activity of sodium ions in the exchange reactions with soil. (DVWK 1983)

[17] T.C. Ministry of Energy and Natural Resources, General Directorate for State Hydraulic Works 1980, p.30; Consortium TAHAL-ECI 1968, D.12

[18] See Köy Hizmetleri 1990

[19] Groundwater is divided into two categories, phreatic (or free) and artesian (or confined). In the Lower Seyhan Plain, the aquifers are generally confined, but they are slightly permeable, allowing slow upward seepage of the confined water. This results in the creation of a virtual water table at a depth which depends on the balance between evaporation, transpiration, rainfall and the upward seepage. This is not a real phreatic water table, and therefore the term 'superficial' water table is used. Superficial groundwater is labelled in Turkish 'taban suyu' and artesian water is 'yer alti suyu'. (see Consortium TAHAL-ECI 1968, C, D)

[20] See Tekinel et al. 1985
The annual and seasonal distribution of precipitation is shown in the Fig. 3.2. and 3.3.

occurs throughout the delta. Pondage is extreme on the highly saline low lying flats near the Sea. Spring rains also recharge the groundwater to some extent. Summer rainfalls rarely affect the water tables; if at all, they replenish the soil moisture depleted through evapotranspiration. The rivers Seyhan and Berdan are water-losing streams. Intrusion of waters from the River Seyhan is only significant near the city of Adana, where the gravel river bed intersects a large shallow continous aquifer which extends to the east of the river. The River Ceyhan is a gaining stream; its influence on water tables is negligible.[21]

The natural conditions do not provide sufficient natural drainage for precipitation and accumulated high standing groundwater; the flat topography, impermeable soil layers, and the lack of natural outlets to the Mediterranean Sea favour rising groundwater tables and, together with the hot climate, the salinization of soils. The depth of the superficial groundwater levels and its salinity content, as well as saline and alkaline soils were first analysed in 1955/56 in the Yüregir Plain, the eastern part of the Lower Seyhan Plain. The EC_e-values varied greatly between extremes: the electrical conductivity value was low in the north, and became progressively higher to the south. In the delta, it varied between 10,000 and 95,000 micromhos/cm, with associated pH-values of 8.20 and 7.55; the percentage of sodium was 91.2 and 47.1 respectively. Particularly high conductivity values of EC_e 130,000 micromhos/cm occurred in the area between the mouths of the rivers Ceyhan and Seyhan, in an area with poor drainage, artesian pressure, and shallow confining cover. The area bordering the Akyatan Lake is relatively saline.[22] Nunns estimated that from approximately 100,000 ha in the lower plain, about 61,000 ha have salinity and alkalinity problems, and 30,541 ha are highly saline.[23] In 1959, a soil study provided more detailed data: it indicated that 19,982 ha were slightly saline (EC_e 4,000-8,000 micromhos/cm), 29,053 ha moderately saline (EC_e 8,000-12,000 micromhos/cm), and 56,602 ha highly saline (EC_e >12,000 micromhos/cm).[24] Most of these lands were found unsuitable for cultivation without improvements, and an artificial drainage system would have been necessary even for improving rainfed agriculture.

---

[21] Data on artesian aquifers which might have an effect on the superficial groundwater tables, are not available.

[22] Drenaj Etüdleri 1957, p.46; Consortium TAHAL-ECI 1968, D-20

[23] See Nunns 1956
A groundwater table observation programme was installed by the Tarsus Irrigated Farming Research Institute in 1952. In 1956, the Regional Directorate for State Hydraulic Works started its first monitoring programme with the installation of 78 observation wells, and, in 1988, 880 groundwater observation wells had been installed within the project area.

[24] Nunns 1956; Dinc et al. 1986
A saline soil is one in which the conductivity of the saturation extract (EC_e) is greater than 4,000 micromhos/cm (25°C) and the sodium percentage of the exchangeable adsorbed cations on the clay complex is less than 15. The pH-value of the saturated soils is usually less than 8.5.

2.

The State's Responsibility for the Development of Water Resources

In Turkey, the basic principles governing water resources are that water is a public good, and subject to rights of prior use. Apart from privately owned springs, surface and ground-water resources cannot be owned, but they are subject to user rights which can neither be sold nor transferred. Water user rights are limited to beneficial uses, such as domestic and agricultural use, fishing, hydropower generation, industry and mining, transportation, and medicinal and thermal uses. Laws regulate public sector activities, e.g., by defining the responsibility for the construction of water networks, the operation and maintenance obligations, and their financing. In the field of rice cultivation, a particular provision is made for Rice Commissions who determine the extent and location of rice fields, and supervise irrigational operations therein.[25] Special legislations on the harmful effects of water have been enacted, e.g., flood control, drainage and sewerage, and on water quality and pollution control.[26] The General Directorate for State Hydraulic Works (DSI) is responsible for the preparation of master plans for each river basin, and establishes priorities for the development of water resources, e.g., the implementation of flood control, irrigation systems, land reclamation and power generation. Priority projects are submitted to the State Planning Organization for their incorporation into the five-year development plans and into the annual programmes, and they should proceed the long-term develoment plans for economic, social and cultural development.[27]

Large-scale public irrigation projects are jointly undertaken by the General Directorate for State Hydraulic Works and the General Directorate for Soil and Water (TOPRAKSU): the DSI is responsible for the construction of the major engineering works, such as the storage facilities, the diversion weirs, the main irrigation and drainage channels and the tertiaries, and the TOPRAKSU is responsible for the on-farm development activities, such as land levelling, the installation of field ditches and field drains, etc. In other words, the DSI controls the irrigation water, while the TOPRAKSU is concerned with the use of water on land and the development of irrigated agriculture. Technical assistance and services to the farmers are provided by the TOPRAKSU and by the Extension Service, and financial assistance comes from the Agricultural Bank.

When the implementation process started in 1963, a coordination committee was established on the central and local administration level, including all involved ministries (see Fig. 3.4.). The Extension Service under the Ministry of Agriculture

---

[25] See FAO 1975, Vol.1
[26] See FAO 1975, Vol.1
In 1988, the Water Pollution Control Regulation was passed by parliament, which identifies water quality classes, protection zones, and various effluent standards for industries. (Law No.19919, 1988; Regulation No.20106, 1989)
[27] See Sayin et al. 1967; Onat et al. 1967
The State Planning Organization is attached to the Prime Minister's office.

was to give advice to farmers on irrigation methods, crop patterns and their diversification, the use of modern inputs, and pest control, with the assistance of a cotton research centre and a plant protection institute.

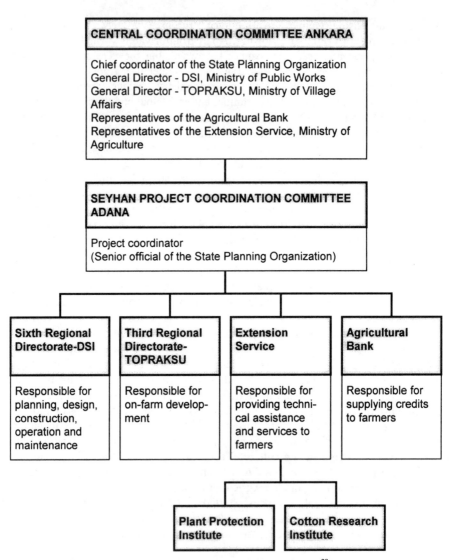

Fig. 3.4. Organizational Chart for the Lower Seyhan Irrigation Project[28]

[28] Source: The World Bank 1978, Chart II

2.1
The General Directorate for State Hydraulic Works

In December 1953, the DSI was established under Law No.6200, and the Sixth Regional Directorate was inaugurated in March 1954 in the province of Adana.[29] The General Directorate for State Hydraulic Works is attached to the Ministry of Public Works. The DSI is a legal entity with an annexed budget.[30] Law No.6200 defines the DSI as the main state agency for developing surface and ground-water resources, to make the optimal use and to develop them in a way to achieve the optimum benefit. Its duties and responsibilities include the construction of protective facilities for flood and torrent hazards; the construction of irrigation and drainage networks; the drainage of swamp areas; the construction of hydropower generation facilities; the construction of water supply and sewage facilities for cities and villages,[31] and the operation and maintenance of the facilities. At a national level, the DSI coordinates water use of the other agencies, which must obtain prior approval from the DSI for each project which is to be undertaken. As the licensing authority, it needs to approve both the use and the extraction rate of water for different purposes.[32]

The Sixth Regional Directorate in Adana manages the catchment area Dogu Akdeniz with the rivers Seyhan, Ceyhan, Berdan, Göksu, and Asi in the three provinces of Adana, Hatay and Icel. The regional directorates are organized along a horizontal and a vertical division, and they, together with field branches, are responsible for the construction, and the operation and maintenance of public infrastructure projects, together with the carrying out of surveys and monitoring activities. In the project's beginning, two sections of the Sixth Regional Directorate were entrusted with its realization: the 'Office of Seyhan' and the '61. Chief Engineering Department' which were later united to form the 'Seyhan Plain Control Office'. Field offices were established on the west and east sides of the River Seyhan to handle project maintenance and coordinate operation, with suboffices to provide contact with the farmers, and command water distribution. Seventeen engineers and 40 assistant personnel worked in the Seyhan Plain Control Office. Key employees had received training in the United States with the Bureau of Reclamation. There was, however, a shortage of engineers in Turkey. Considerable

---

[29] In 1880, a commission was appointed under the Ministry of Public Works for water resource development and, in particular, for the development of irrigation and drainage projects. Between 1905 and 1913, the Commission implemented only one large irrigation project in Central Anatolia (Konya). After the foundation of the Turkish Republic in 1923, a Directorate for Water Works within the Ministry of Public Works dealt with drinking water supplies for the capital and the drainage of swamp areas in Tarsus and Aynaz. In 1951, a so-called "Great Water Congress" initiated and promoted the establishment of an effective organization for the development of water resources. (see Akalin 1966)

[30] The organization and the general framework of procedures were patterned after the U.S. Bureau of Reclamation.

[31] Since 1968, the DSI has to supply drinking water, utility and industrial water for cities with a population of more than 100,000. (Law No.1053)

[32] See Bilen and Uskay 1992

competition existed between the Turkish government and the private industry, and since salary rates favoured private industries, the Government often lost, and continues to lose, personnel.

2.2
The General Directorate for Soil and Water

Law No.7457 by which the General Directorate for Soil and Water was established came into force in February 1960. The Third Regional Directorate, to which Adana belongs, was newly established when the implementation process began in 1963. The General Directorate for Soil and Water was at first assigned to the Ministry of Village Affairs, but was then, in 1985, transferred to the Ministry of Agriculture, Forestry and Rural Affairs.[33] It is affiliated to the ministry, with an annexed budget. The agency is concerned with soil surveys and soil conservation, the construction of minor irrigation schemes which are turned over to other entities after construction (e.g., smaller than 500 l/sec.), and on-farm development activities. According to Law No.3202, the TOPRAKSU should „utilize the water and land resources in the best way available for agriculture, and should carry out the necessary activities in relation to the conservation of these resources".[34]

In large-scale public irrigation schemes, the TOPRAKSU is responsible for all on-farm development activities which are necessary to use the water obtained from state-constructed irrigation facilities, such as land levelling, the construction of field-side channels, and on-farm irrigation and drainage facilities.[35] It promotes the foundation of farmers' associations and cooperatives to support the management activities related to irrigation, soil conservation and land development operations. The regional directorates of TOPRAKSU are, like those of the DSI, organized along horizontal and vertical divisions. The Lower Seyhan Project lies within the jurisdiction of the Third Regional Directorate which comprises the provinces of Adana, Icel, Hatay, and Gaziantep. The offices involved alongside TOPRAKSU in activities for the Lower Seyhan Irrigation Project were the South Regional Directorate, the Adana District Development Section, the Seyhan Project Office, the Tarsus Irrigated Agriculture Research Institute, the Topraksu Land Conservation Station, the Alifaki Saline Soil Reclamation Substation, the Tarsus Training Center, two units carrying out construction contracts for farmers with credits provided by the Agricultural Bank, and some demonstration farms. TOPRAKSU was the state agency with the largest responsibility for the development programme.

---

[33] See MAFRA 1987. In 1985, the General Directorate for Soil and Water (TOPRAKSU) was reorganized as a General Directorate for Rural Services (GDRS) under the Ministry of Agriculture, Forestry and Rural Affairs. (Law No.3202) The abbreviations 'TOPRAKSU' and 'GDRS' label the same organization.

[34] Sayin and Yegin 1966, p.237

[35] Law No.3202, Article 2, j

3.
The Project Planning and Implementation Process, and its Evaluation with Regard to High Groundwater Levels and Salinity

The Lower Seyhan Irrigation Project area comprises 175,000 ha which were to be installed with irrigation and drainage facilities and with on-farm development works. The Turkish state agencies responsible for planning and implementation estimated a construction period of eight years, from 1961 to 1968, for the DSI related works, and a period of ten years for TOPRAKSU's works. Both schedules were, however, subject to continous revisions, and were first changed in 1963 by the IBRD/IDA, when subprojects of varying sizes were also defined (see Table 3.1.). The Lower Seyhan Plain, however, was to be developed in three (later in four) stages in the designated areas (see Map 3.2.).

Table 3.1. Implementation schedule for the subprojects, and their sizes

Stage I	65,000 ha	
	DSI	1963 - 1968
	TOPRAKSU	1966 - 1978
Stage II	48,600 ha	
	DSI	1969 - 1977
	TOPRAKSU	1970 - 1980
Stage III	19,830 ha	
	DSI	1976 - 1986
	TOPRAKSU	1981 - 1987
Stage IV	40,657 ha	
	DSI	TP (1991), YP (not yet started)
	TOPRAKSU	not yet started

In 1987, with the completion of the stage III project, irrigation and drainage facilities had been constructed on approximately 133,000 ha, or about 70% of the total project area, with 69,031 ha in the Tarsus Plain and 64,400 ha in the Yüregir Plain. Surface drains were installed on 120,000 ha, close conduit drains on 62,000 ha, and land levelling had been completed on 95,400 ha. According to the DSI's and TOPRAKSU's estimation, the stage II and stage III projects were completed with delays of 10 years and 18 years respectively.[36] The service area was not significantly extended until the 1990s.

---

[36] FAO 1988; DSI, 6.Bölge Müdürlügü 1986/87

Map 3.2. Boundaries of the subprojects[37]

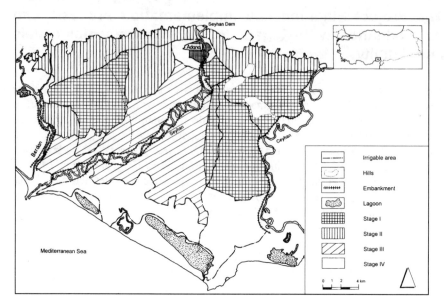

3.1
High Groundwater Levels and Salinity as Reported in the Basic Planning Documents

The Lower Seyhan Project Report was the basic document prepared by the General Directorate for State Hydraulic Works and the General Directorate for Soil and Water to apply for loans/credits from the International Bank for Reconstruction and Development/International Development Association.[38] The report was based on previous studies and surveys.[39] The Comprehensive Plan of Development for Irrigation and Drainage of the Adana Plain, made by the engineering company Tippetts-Abbett-McCarthy-Stratton (TAMS), analyzed the soil and drainage conditions and the necessary requirements to develop a plan for the irrigation and drainage infrastructure facilities before the commencement of construction. Its financing was part of the first IDA loan (63-TU; US$ 400,000). The Lower Seyhan Project Report, i.e., the basic planning document, emphasized the problems of high groundwater levels and salinized soils as being of major importance (see Box 3.1.).

---

[37] Source: DSI (undated)

[38] DSI and TOPRAKSU 1961

[39] Report on Economic Study of the Lower Seyhan Valley 1949; Economic Report on the Comprehensive Seyhan River Development 1951; Comprehensive Plan of Development for Irrigation and Drainage of the Adana Plain 1961

Box 3.1. High groundwater levels and salinity as reported in the Lower Seyhan Project Report[40]

The high water table in the Seyhan Plain is the main problem. It keeps the agricultural production at a low level and even under the present conditions (without the implementation of the new project), it creates a need for drainage and improvement. It is evident that when irrigation starts, the importance of this problem will increase and it will become necessary to take appropriate measures.

The water table in the Seyhan Plain is generally at a depth between 0.5 and 5 meters. Upon closer observation it is seen that some parts of the plain present special conditions. For example, the water table in the areas below the 2.5 meter contour is always high. During the winter months, in greater part of the areas, rain water accumulated by surface drainage remain on the ground in the form of lakes, and at such times the surface waters and the groundwaters show a tendency to join. This condition can be best observed in the area beginning near the Yemisli Village in the south of the Yüregir Plain and extending as far as the hills near Karatas. The depth of water in this artificial lake is about 40 to 45 centimeters. These waters flow to the lagoons, and those waters from surface runoff which do not find an outlet to the sea dissappear through evaporation when the weather becomes warmer. Even in the summer months the water table in this area is at a depth of 0.8-1.2 meters.

As a result (of the surveys and investigations) it was found that out of a total irrigable area of 186,244 ha only 3,336 ha do not require drainage and improvement. In an area of about 87,000 ha, which does not need improvement or needs only slight improvement at the present, drainage will be required when irrigation is developed. It will be necessary to construct deep drains in 14,000 ha in this area. An area of about 96,000 ha needs drainage before the development of irrigation.

Drainage is needed in almost every part of the plain. It is possible to meet this need by the project accompanied by laying farm drains as required. It is proposed that the soils be improved by leaching during the irrigation and for carbon dioxide producing crops such as rice and Bermuda Grass to be planted.

More than half of the land in the project area already had drainage and salinity problems producing negative impacts on yields. Large funds were required to apply effective improvement measures. The IBRD/IDA report underlined the lack of credits available for the farmers already irrigating to built on-farm drains, and stressed the deficiency of main drainage systems.[41] Drainage works ranged on first order of the project outline:

> The project involves drainage by gravity for the part of the plain above the 2.5 meters contour and drainage by pumping for the part of the plain below this contour which covers 36,000 ha. It will be required to build 3 pumping stations for this purpose (...) It is necessary to take improvement measures on 96,000 ha of the project area to solve the alkalinity and salinity problems.[42]

[40] Source: DSI and TOPRAKSU 1961, p.19
[41] Ibid., pp.24-25
[42] Ibid., p.36

Pumping stations would be necessary in the southern part of the plain due to the lack of natural outlets to the Mediterranean Sea.[43] The Department of Technical Operations of the IBRD/IDA approved the project in 1963 as sound and suitable for an IDA credit, although it would be expensive due to the need of an extensive drainage system.[44] The IBRD/IDA team repeatedly underlined the importance of drainage works, and detailed the technical design of the existing and proposed gravity drainage system.[45] The report emphasized that „drainage is the most important physical factor in developing the plain for irrigation. All of the plain must have a drainage system to remove surface water resulting primarily from heavy rainfall during the rainy season. Approximately 85% of the plain would require subsurface drainage to control the water table".[46] (see Box 3.2.) The team agreed with the DSI and TOPRAKSU to which extent they would be required, the spatial distribution and the technical solutions. None of the very problematic areas were included within the first project under consideration for IDA financing, which lays in the northern section of the plain with soils which are more easily drained. However, the basic planning documents of the Turkish agencies and the IBRD/IDA team portray that the problems of high groundwater levels and salinity were, thus far, acknowledged, and the subprojects were issued with technical measures for their alleviation.

Box 3.2. Drainage requirements as reported in the IBRD/IDA document[47]

In the northern part of the plain, where gradients are satisfactory, surface drainage is not a problem. Nevertheless, drainage is required to remove the surface flows from the project quickly and without damage to lower areas. (...) Generally, the texture of the soil profile becomes progressively heavier and less permeable from north to south. The amount of drainage required would be in the same relation; that is, proceeding from the higher to the lower elevations more drainage with generally closer spacings would be required. The central and lower delta areas have high water tables (...) Approximately 20% of the plain bordering the Mediterranean Sea and lying below the 2.5 meter contour line is too low for gravity out-fall drains and pumping would be required to evacuate the drainage water which could be conveyed to the pumps by open drains.

---

[43] Ibid., p.35
[44] IBRD/IDA 1963, p.20
[45] Ibid., see Annex 1, p.3
[46] Ibid., p.4
[47] Source: IBRD/IDA 1963, p.4

3.2
The Implementation and Evaluation of the Engineering Work

The implementation of the subprojects I and II is well documented in reports from the IBRD/IDA and the Turkish authorities. The documents are reviewed with regard to constraints and shortcomings on groundwater table and salinity control means. Whenever data are available from IBRD/IDA and the Turkish authorities, both views are presented. After the construction work had been finished, operation started immediately, and thus, some of the problems mentioned relate to the operation of the system rather than to the implementation of engineering works but they are, nevertheless, reported.[48]

3.2.1
The Stage I Project and its Shortcomings

The main tasks in the stage I project area included the modification and rehabilitation of some of the previously constructed works. In the 1940s, the Seyhan right bank irrigation system was planned to serve 16,500 ha of irrigable land, but an approximate average of only 3,300 ha could be irrigated. The system had been poorly designed, and could not deliver adequate water to the area. Its maintenance had been neglected, and an inadequate drainage system had been provided. The stage I project would correct these deficiencies, and comprised, further, of the completion of constructions which were already in progress, i.e., laterals, tertiaries and drains, to serve a gross area of 65,000 ha in the subdistricts 1-TP, 1-YP, 3-YP and 4-YP.[49] The estimated expenditures for the stage I project was expected to be US$ 50.1 million; the amount of the proposed IDA-credit was US$ 20 million, or about 40% of the total estimated costs. 40% of the total project costs were designated for on-farm development works, 4% for extension services and 54% for construction. Within the stage I project the Regional Directorates of the DSI and TOPRAKSU would provide the listed works in the Boxes 3.3. and 3.4. below:[50]

The DSI's construction works would be carried out by contractors, and by the DSI's own forces at the expense of the annual budget, and on-farm development works by TOPRAKSU, or by the farmers under TOPRAKSU's supervision. The DSI's works would be implemented between 1963/64 and 1966, and between 1963 and 1968, 80% of TOPRAKSU's work would be carried out; about 20% of its work was then left to be done after 1968. It was expected that after the project reached full development in 1973, the DSI would levy irrigation charges sufficient to cover the annual costs of operation and maintenance, and a capital charge to recover all the investments made in the irrigation works since 1953. The Government proposed to set irrigation charges high enough to recover, at least, O&M costs.

---

[48] The operation and maintenance of the system is analyzed in Chapter Four.
[49] The abbreviations YP and TP stand for Yüregir Plain and Tarsus Plain.
[50] See IBRD/IDA 1963; Balaban 1972

Box 3.3. Sixth Regional Directorate for State Hydraulic Works - Stage I Project

---

(1) Modification of the irrigation and drainage system built in 1946 to irrigate about 16,500 ha;
(2) construction of irrigation and drainage networks to serve an additional irrigable area of 36,500 ha;
(3) construction of the eastermost main drain to the sea;
(4) modification and rehabilitation of the diversion dam, including the installation of control gates, and the re-shaping and lining of the 40 km right bank (54 m³/s) and the 19 km (90 m³/s) left bank conveyors to serve 17,500 ha and 1,000 ha respectively.

(3) and (4) would also serve stages II and III.
    Construction of tertiary laterals, drains and operating roads to enable 53,000 ha (net) to come into operation.

Box 3.4. Third Regional Directorate for Soil and Water - Stage I Project

---

On-farm works on 65,000 ha (e.g., closed drainage collectors, excavation and field roads), land levelling on 26,575 ha, on-farm drains and on-farm ditches on 23,679 ha. The reclamation involved leaching of the salt from highly saline lands.
    Farm drains should have a spacing of 50 meters and a depth of 1.8 meters in heavier soils, a wider range and 1.2 meters depth in soils with better permeability, and in areas where the groundwater was below 3 meters, drainage was deferred until the groundwater level rose.

Amortization would be over a period of 50 years, beginning in 1973, after completion of the 10 year development period from 1963 to 1972, because it was assumed that the farmers would not be able to afford to pay irrigation charges high enough to enable irrigation investments to be recovered with interest until they had completed their repayments of loans for on-farm development.[51]

3.2.1.1
Evaluation by the IBRD/IDA

By 1965, some 30,000 ha were under irrigation, partly from project installations. In 1966, with the stage I project nearing completion, an additional 30,000 ha were not irrigable.[52] In 1966, three years after commencement, a team from the IBRD/IDA evaluated the ongoing project in regard to drainage and levelling needs, irrigation practices, and the requirements for the improvement of irrigation efficiency in the stage I area:

Overrun of construction costs and changes in scheduling
The implementation of the stage I project was delayed, and led to cost increases of 11.6% over original estimates. The original cost estimates were TL 486.1 million compared to the revised cost of TL 544.0 million. The internal rate of return was

---

[51] IBRD/IDA 1963, p.18
[52] See IBRD 1966

not 10.7%, as expected, but 8.7%. The changes in the anticipated benefits originated in delayed extension activities and a critical shortage of short-term credits.

Civil works were completed in 1968, 2.5 years behind schedule, and land improvement originally scheduled for completion in 1970 was finished in 1973. TOPRAKSU did not start the on-farm drainage programme until 1966, and the project I area was first provided with on-farm drainage systems in 1978.[53] The IBRD evaluation team assumed that the delays resulted primarily from the relative inexperience of the DSI and TOPRAKSU. The DSI experienced difficulties in placing contracts for major works and purchasing its equipment. Complex contract procedures and inadequate budget allocations extended the execution of contracts already under way. Expenditures on construction up to March 1965 was approximately TL 56.5 million, compared with the original estimate of TL 162.0 million. TOPRAKSU had to overcome similar problems. A particular problem which caused the delays of TOPRAKSU's works and aggravated the inability to carry out on-farm improvements in the first years, referred to the farmers' reluctance to accept a heavy credit burden for financing on-farm works; in addition, the Agricultural Bank could not provide sufficient funds for long-term loans, and the farmers would not borrow from the private sector which had high unsubsidized interest rates of 70% per annum.[54] Up to March 1966, expenditure on land development was TL 22.9 million, compared with a planned amount of TL 99.0 million. In August 1966, the organizational set-up of the project was revised, and TOPRAKSU charged on-farm improvements directly to its own budget, planning to recuperate the cost at a later stage through increased water charges.[55]

Inadequate land levelling and drainage
Only a very small area had been accurately levelled. Due to wrong ploughing practices depressions had formed. On the flat slopes, and under existing conditions, the irrigation stream could cover 80 and 90% of the land. However, part of the land received too much water, and some parts less than the crops needed, resulting in overall field irrigation efficiencies of less than 30%. Ponding in fields caused plant damage, and water was wasted from percolation and surface runoff. A salinity built up, or a rise in the water table resulted, and light and medium degree land levelling had to be carried out on 92% of the area of stage I.

The nearly flat surface requires facilities for the removal of surface water, whereas the percolation of winter rainfall and the application of irrigation water call for a system of deep subsurface drainage for salinity and water table control.[56] Surface drainage could be obtained through a combination of land levelling and shallow field ditches which removed the waste water to the main drainage network. Accurate levelling was by far the most important of the two measures. The main drainage system had been designed to have sufficient capacity to accomodate both surface and subsurface requirements. The depth specified in the design for the

---

[53] See Köy Hizmetleri, 3.Bölge Müdürlügü 1992, 1993, 1995
[54] See The World Bank 1985, p.39
[55] IBRD/IDA 1969, Annex 7, p.2
[56] Ibid., p.8

outlet of the deep drains assured that these drains would be capable of handling all surface drainage waters.

The most serious drainage need was the control of the groundwater level beneath the irrigated fields through subsurface drainage, because without this the groundwater table would rise closer to the surface and aggravate the salinity situation. Collector drains less than 1.5m in depth were insufficient. 1.6m would be the minimum drain depth for arid land, while a depth of 1.8m to 2.0m would be preferable. The IBRD-team recommended priority areas for drainage: first priority should be given to areas with poor subsurface drainage (10,780 ha); second priority was designated to areas (9,220 ha) with underlying saline groundwater greater than EC_{gw} 4,000 micromhos/cm.[57] The recommended spacing for first priority areas in stage I was 120 meters for initial tile drain installations;[58] the areas with second priority should have spaces of 150 meters due to the higher permeability of soils. A subsequent installation should provide tile drains midway between the original installations in those areas that required them to be more closely spaced. These installations should be carried out by the farmers. The Evaluation Team assumed that the farmers' acceptance of drainage practices was low not only because of the heavy expenses, but also because of their unawareness of the necessity for drainage. The farmers would not be aware of the benefits that accrue from controlling the water table and salinity.

The executed land levelling and drainage programmes were uncoordinated: the installation of drainage prior to land levelling resulted in damage to the drains when the necessary levelling was carried out. The DSI-constructed main drainage system was not always of adequate depth for the TOPRAKSU tile drain collectors. When the adequate depth - at least 2.5m - was not available, TOPRAKSU should have delayed installation of the collector drains until the DSI open drains could be deepened.

Insufficient and unmeasured water deliveries

The tertiary canals in the subunit 1-TP did not deliver water to the farms, and quarternary canals were necessary. Of 53,000 ha net irrigable land, 10,185 ha, or 21%, were not served with irrigation water. The land ownership pattern complicated the water distribution, because the deliveries did not follow the actual property lines. The unserved farmer had to make provisions for the conveyance of water across neighbouring plots. It was customary for the farmer whose land was being crossed to use the conveyance ditch for his own supply needs. Thus, the downstream farmers were affected by upstream water use, and there was in general a severe water shortage at the lower end of the ditch. Serious delivery shortages occurred, and resulted in a low water application efficiency of 30% at on-farm level. The project was planned on the assumption that an on-farm application efficiency of 70%, or higher, could be reached.[59] Orifices from the main to the secon-

---

[57] See IBRD 1966, p.11, Table II

[58] Tile drains are closed drains which are composed of pipes made of sections of hollow earthenware or concrete tiles that are buried at a depth of 1 to 2 meters. Excess water in the soil seeps into the pipes through apertures in the tiles. (Landis 1994)

[59] See IBRD 1966, p.7

dary, and from the secondary to the tertiary channels had not been calibrated.[60] No plans had been made for measuring the supply of water from the tertiary channels to the farms. The team recommended siphons.[61]

3.2.1.2
Evaluation by the Turkish Agencies

In 1967, the General Directorate for State Hydraulic Works conducted a survey of government-financed and operated irrigation schemes of which the Lower Seyhan Irrigation Project was the largest.[62] Only 54% of lands that should have been irrigated according to the project plans could be irrigated with irrigation systems which had been constructed by the state. The remaining 46% could not be irrigated due to the fact that irrigation canals and related water control structures had not been completed. Irrigation canals were not designed properly to fit the local conditions, e.g., irrigation water had to be pumped to the field, because some delivery channels were below field level. The installation of canalettes after 1965 solved this problem.[63] Some of the farms, or parcels of land, did not have access to irrigation water due to the small, scattered and irregular landownership pattern, and because the distance between the tertiary channels was too great (i.e., 500-600m). In some parts access to the delivery network was difficult, and the neighbouring farmers refused right-of-access. Excessive application of water, and only 18 hours irrigation in a day, was creating an artificial water shortage, while in other places farmers, being unaware of the benefits of irrigation, were not willing to irrigate their land due to the extra expenses. In other parts, drainage water was stored, first, because the farmers closed the installed drains for rice cultivation, and second, because the levels of the drainage system were unadjusted between the DSI's and TOPRAKSU's executed works.[64]

The lands were not prepared for irrigation through land levelling and field drainage systems. From the reviewed state projects, 53% of lands within the systems required light land levelling, 29% medium, and 8% heavy levelling; 40% of these lands needed surface drainage, 26% tile drainage; 73% of the total were irrigated through unorthodox flooding methods, causing water and soil losses, and only 10% were irrigated by means of furrows. Irrigation water was not applied at the proper time, in proper amounts, and with proper methods. Wastage of water occurred, because farm irrigation structures and measurement devices did not existent. The yields were below the estimated level due to the insufficient use of fertilizer and low yield varieties (seed), and the lack of plant protection.[65]

---

[60] Orifices are holes or openings, usually in a plate, wall, or partition, through which water flows, generally for purpose of control or measurement.
[61] See IBRD 1966, p.7
[62] Only points in the report which refer to the Lower Seyhan Irrigation Project will be mentioned.
[63] Canalettes are precast elevated cement shells.
[64] See Uner 1967, p. 232; FAO 1987; DSI, 6.Bölge Müdürlügü 1986/87
[65] See FAO 1987; DSI, 6.Bölge Müdürlügü 1986/87
In the 1960s and 1970s, fertilizer and pesticides were not available to the required levels.

Constraints on technical solutions, on improving irrigation methods and on water saving

At the time of the report, almost all the lands under irrigation were irrigated by gravitational methods. Other irrigation technologies, such as sprinkler, were not installed because of the large investments necessary, the high price of fuel, and the unavailability of electricity for the sprinklers. The report advocated sprinkler irrigation because it would save water, and eliminate weeds, erosion and siltation problems in the delivery network; it would minimize the need for drainage and the problems associated with rights-of-way, and, most important of all, would shorten the development period. Expensive and time-consuming land levelling would not be necessary, and the farmers could be easily taught how to operate it. Thus, it was recommended that every effort should be made to extend sprinkler irrigation in areas where it would be economically viable.[66]

Because of the incomplete delivery systems, improper alignment, grade, elevation and the capacity of the irrigation canals, the lack of water measuring devices for delivering water to farms, only approximately half of the service area was actually irrigated. Water measuring devices were not included in the new projects, and the unmeasured water delivery made it difficult to discourage the wastage of water and to encourage the farmers to level their lands to save water. The fact that lands were fragmented and scattered was another reason why not all areas could be irrigated. Some parcels did not have access to water, and the cost of irrigation development was high, because additional canals, drains and roads were required to provide access to the isolated parcels in the service area of the projects.[67] The inheritance law in Turkey promotes land division, making the task of irrigation development more difficult. It was assumed that the farmers would not understand the benefits of on-farm development activities, and, because of their conservative attitudes, it would not be easy to teach them improved irrigation methods.[68]

According to the legal regulations, the water charges for O&M have been determined on an acreage basis without considering the volume of water used by the farmer, and the farmer has not to pay for his wasted water. The overuse of water, which created water shortage and drainage problems, was common in irrigation systems built by the state. Thus, water saving as one of the benefits of land levelling did not make sense to farmers in Turkey.

---

[66] See Uner 1967
[67] Ibid., p.237
[68] Ibid., p.239

Considerations on investments

The investments made in the agricultural sector were far from sufficient, and it was impossible to provide irrigation over the entirety of the designated service area. Financing activities should have been obtained through the investment of the public and the private sector. However, the responsibilities of the public sector were not clearly defined, and measures were not taken to create the necessary conditions for private sector involvement. On-farm development activities such as land levelling, drainage and farm irrigation systems were only beneficial for the farmers, and for this reason, it was thought that they should be financed by the farmers either with their own money, or with loans made available to them. The authors consider of the fact that investments in water-saving means could result in national benefits in terms of saving water, preventing waterlogging, and improving saline and alkaline lands, on-farm development would have to be financed by the state if the farmers themselves were not able to do so.

Most of the farmers in Turkey, however, were not able to finance on-farm development activities with their own money. Thus it was assumed necessary to provide low interest, long-term loans. Two different funds, namely the TOPRAKSU fund and the Agricultural Bank's own funds, were made available for these loans. Funds, however, were not sufficient, and the majority of farmers could not make use of those available, because lands were assessed at values too low to provide sufficient security for the loan; there were deficits in cadastre-registration; disputed lands were not accepted as security (no matter how small the portion of disputed land was). Between 1947-1964, nationwide, 1.8 million ha of state land had been distributed to 361,291 landless farmers. Under the laws relating to the distribution of these lands, the title deeds were only to be given to their new owners 10 or 25 years after the date of distribution. The farmer whose lands had no title deeds could not get loans. The Agricultural Bank would also not accept the joint liability of farmers for long-term loans. To overcome financial constraints, the farmers could use TOPRAKSU's equipment at low prices. Even if sufficient loans could have been given to the farmers, or land levelling and drainage would have been financed by the state, the lack of necessary equipment in Turkey would have prevented the implementation of programmes to the required capacity. Thus sufficient foreign exchange would have to be provided to import the necessary equipment, or the private industrial sector encouraged with the necessary loans to manufacture this equipment in Turkey.[69] Some of the farmers with large farms were reluctant to irrigate their land and to practice intensified farming methods, because their incomes were already adequate enough for their needs. These farmers did not realize that they could get the necessary money for on-farm development which would increase their overall profit by selling off a small portion of their land.[70] Some of the irrigated lands were cultivated by tenants. On such lands, neither the landlord nor the tenant made any investment for on-farm development.

---

[69] Ibid., p.238
[70] Ibid., pp.238-239

Administrative factors
The different activities related to irrigation development were also separately planned and programmed, and this caused the dispersion of limited funds, equipment, and personnel over the country. The lack of farmer associations and water districts made it necessary for the government agencies to deal with each farmer separately. This wasted money and time.[71]

3.2.2
Major Changes Within the Implementation Period of the Stage I Project

Within the implementation period of the stage I project major changes were initiated which refer to the investment strategy, to the establishment of a farmers' extension service and to changes in the management structure.

3.2.2.1
The Investment Strategy

In 1963, it was politically decided that the farmers in the project area should finance the investments for on-farm development works from their own resources or through credits provided by the Agricultural Bank, and TOPRAKSU would execute these works. TOPRAKSU's equipment could be rented out to the farmers at low prices. This was significantly different to the DSI-related works; its investments were centrally financed and executed, and, in the beginning, reimbursment from the beneficiaries was not foreseen.

The Agricultural Bank was the major source for short and long-term development and production credits which could be provided directly to the farmers or through their cooperatives.[72] Credits would only be provided on the basis of farm plans, prepared and approved by TOPRAKSU, with a repayment period varying from 5 to 20 years, bearing interest at 5%/a; the Bank would keep separate records and accounts; the Bank would also allocate sufficient staff to prompt credit applications, and for the supervision of loans and the collection of payments.[73] Any farmer could apply for a land improvement loan (e.g., for land levelling) from the Agriculture Bank. The Bank's approval of a loan depended on an assessment by TOPRAKSU of the proposed improvements. Final approval was given by a joint committee of the Bank and TOPRAKSU. The committees met weekly, because, according to the law, an application had to be processed within 45 days. The local branch of the bank was authorized to extend the credit after the loan had been approved, and it then notified the farmer and TOPRAKSU. The actual implementation of a project was then monitored by TOPRAKSU. The value of the project, rather than the wealth of the farmer, was in theory decisive. Less affluent farmers, however, faced some handicaps because 10% of the cost of the project had to be

---

[71] Ibid., p.239
[72] The T.C. Agricultural Bank, established in 1888 and reorganized in 1937, was given broad responsibility to finance agricultural development and related industries.
[73] IBRD/IDA 1963, p.17

put up by them, and the Bank received the first mortgage.[74] In 1966, in an early stage of the project's implementation, this regulation was changed, and, to date, the TOPRAKSU's investments in on-farm development works were completely subsidized.

When the implementation of the first subproject started, on-farm development was a new and unproven concept, and TOPRAKSU was a newly established state agency with inexperienced staff. Most of the country-wide available equipment for improving land by levelling, on-farm drainage systems, on-farm channels and furrows had been assigned to the Lower Seyhan Irrigation Project, as had most of the trained engineers and technicians. TOPRAKSU's machinery capacity was increased only for the use within two large-scale state projects, of which the Seyhan Project was one. Both projects received the bulk of TOPRAKSU's investment for irrigated agriculture: Seyhan's share ranged from a low of 23% in 1966 to a high of 62% in 1968. From 1965 to 1971 TOPRAKSU's budget had roughly doubled, and the amount allocated to the Seyhan project was about 45% of the total budget assigned for irrigation development.[75]

The strategy in use for the TOPRAKSU's related works has become known as the Seyhan Strategy, which means that the works to be done were financed through general budget revenues, and were executed by the responsible government agency with its staff, equipment and machinery, because in the project's beginning, there were neither land development contractors nor domestic land-shaping equipment. The Seyhan Strategy was designed to bring rapid development to selected projects, and, in particular, rapid on-farm development in public projects, because there was always a large gap between lands where primary water delivery systems had been completed and the far smaller amount of land for which on-farm works had been ready.[76] Until 1983, the work was carried out by TOPRAKSU with its own equipment, machinery and personnel; after 1984, however, construction was contracted out to the private sector, but the speed of development was still ultimately dependant on the size of the allocated budgets, although the money spent per hectare and the time a private contractor needed was more viable, e.g., the private sector has a far higher equipment utilization rate.

Before 1966, the DSI's investments should have been totally subsidized but this was never effectual. The DSI's establishment Law No.6200 was changed, and determined, backdated, that the reimbursment of the DSI's capital costs should be amortized over a period not exceeding 50 years. The Council of Ministers decides on the length of the repayment period and when it was to begin; repayment usually starts five years after the beginning of the operation phase. A certain amount of interest rate could be applied, which has been subject to approval by the government, but in practice no interest has been charged. Furthermore, capital charges, once they had been established, have not been adapted to match inflation. They are

---

[74] See U.S. A.I.D. 1983
   The interest rate on agricultural loans rose from 8% in 1974 to 19.5% in May 1981, but it was still less than the average inflation rate which was 29.7% per year in the 1970s.
[75] See Mann 1972, p.8
[76] See Pfister 1971, p.79

usually adjusted every five years, and they can only be adjusted with the approval of the prime minister when the charges are very low. The legislation has stipulated that the investment costs are not only to be borne by the beneficiaries of a certain project, but that the investment costs, differing from region to region, are to be balanced among the regions. Beneficiaries with a low income would not have to pay. One third of the investment costs for the Seyhan Dam and the reservoir, and the investment costs for the irrigation networks were assigned to the irrigation sector. Costs which are not repayable were costs for flood control and nature conservation. The second third of the costs was assigned to the energy sector. The landowner has to pay the capital charges annually per hectare for the irrigation and drainage networks.[77] The charges are to be raised per irrigated unit.[78] The same system would be applicable for the repayment of the costs for the drainage networks. The fees would have to be paid by the beneficiaries after the harvest period, between the months November and February, directly to the Ministry of Finance. For delays in payment, a one-off penalty of 10% would be raised.

The political pressure created by the beneficiaries, and the urgency of meeting the needs of the country forced the DSI to construct the irrigation works without making contracts. The DSI did not have the legal obligation to make contracts with the beneficiaries before the construction of irrigation projects, which appeared to be an obstacle for the reimbursment of the state investments even though in accordance with Law No.6200 they should have been reimbursed, at least after 1966/67.

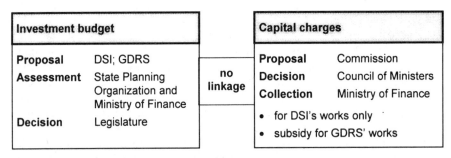

Fig. 3.5. Provision of investments after 1966/67

---

[77] The share of the costs assigned to irrigation is divided by the number of years planned for repayment, and the area of irrigated land, and through this method an amortization value is determined. As irrigated lands are based on estimations, they give rise to distortions. Since the mid-80s indirect charges are being raised, e.g., taxation of agricultural income, of increasing property value and of certain crops.

[78] The nominal investment costs per hectare irrigable area would be higher for the Lower Seyhan Irrigation Project when compared with other irrigation projects. In 1968, they amounted to TL 7,303/ha, compared to average costs of TL 4,626/ha in the same year. (see Pfister 1971, pp.85-90)

3.2.2.2
The Creation of a Farmers' Extension Service

The IBRD/IDA's main concerns with the progress of the stage I project, on which the Turkish Government agreed, were the lack of coordination between the various agencies; the insufficient provision of credits to the farmers; the poor progress of on-farm works, and the need to give more responsibility to TOPRAKSU, since the farmers were unwilling to invest in the improvement works themselves; and finally the inadequate provisions made for agricultural extension.[79] The stage I project was negotiated between the IBRD and the Turkish Government with a commitment to an effective extension service. The government should increase the staff of the Extension Service and provide an adequate number of vehicles in order that appropriate contacts with farmer could be made. It was assumed that the Extension Service would have a great responsibilty in influencing the required change in cropping patterns, advising on irrigation methodes, and pest control, etc.[80] This aim was, however, not achieved. Thus, in 1967, an emergency extension programme was initiated as part of the stage I project.

With the installation of the engineering works and the completion of on-farm development activities, it was expected that the farmers would switch from rainfed to irrigated agriculture, resulting in increased yields and diversified cropping patterns. The prevailing crop pattern with cotton and wheat as the dominant crops would be altered to higher value fruits, vegetables and fodder plants. In the first years, the irrigation efficiency was low; the farmers in the area of the stage I project showed reluctance to crop diversification, and cotton and wheat yields increased little. The lack of extension advice, due to an insufficient extension service, was regarded as one factor contributing to the still prevailing rain-fed agriculture and to the low main crop yields.

The Farmers' Extension Service (CES) was only created for the Lower Seyhan Irrigation Project.[81] The organization was formed from several separate government subdivisions to bring specialists and foremen together in one organization. The director was from the state extension service; specialists were from TOPRAKSU, the Plant Protection and the Animal Husbandry Institutes, and the state extension service. Village foremen were supplied by TOPRAKSU, and village workers were selected from amongst the villagers. The extension system, which became known as the Training and Visit system, was first applied in Turkey, then in many other IBRD/IDA financed projects. In the first year, the service covered 23 villages with local foremen in each village, 4 regional extension specialists and an expatriate extension specialist. The instructors' work was not only based on giving instructions to some 'leader' farmers, but also offering professional aid to others. The regional specialist made regular visits to the villages to give guidance and know-how to the foremen and farmers. The extension service did not reach the planned full coverage of 100 villages in the stage I and II project area, but remained at 86 villages in 1976. Not all the equipment required was procured, and

---

[79] See The World Bank 1978
[80] See IBRD/IDA 1963, p.16
[81] For details, see Aktas 1976

only US$ 106,000 were spent out of US$ 250,000 allocated. After 1976, the Ministries of Agriculture and Village Affairs stopped replacing staff who had resigned.[82]

One important issue of the Farmers' Extension Service was irrigation. The farmers were taught to give the right quantities of water at the right time and with the appropriate methods which would prevent water from being wasted. As a result of the provided service the farmers halted uncontrolled flooding, and changed to irrigation by means of furrows to achieve an even distribution of water. All evaluation reports agreed on the success of the Extension Service with regard to increasing yields of cotton and wheat, to the use of modern inputs and to an increased irrigation efficiency, but also agreed that it failed in getting farmers interested in crop diversification and in changing their traditional practices of land preparation.[83] After 1974, the Extension Service faltered in terms of budgetary support, links to research, and field capability. No new foremen were recruited. The concept of an autonomous service outside the responsible national ministries was later revised. In 1986, it was formally abolished, and its duties were transferred to another extension service (CEY),[84] but some villages still support and finance their CES foremen.

As the extension service did not receive the full support of the central ministries it experienced difficulties, for example, in obtaining necessary equipment, and in achieving full coverage of 100 villages. The Turkish government considered the organization as being a special unit that was not part of any national programme, and had contemplated dissolving it completely with the completion of the World Bank project. Both the central ministries and the government could not see any justification for continuing the allocation of scarce resources and staff to an area which already received priority, and where the work of the special extension service was more or less finished as the farmers had been successfully introduced to irrigated agriculture. There was no one particular agency responsible for supporting the service during the project implementation that would thus be interested in continuing thereafter as it was a special unit incorporating many agencies. The weakness of an organization which was not included in the general government structure, and which received its staff from different ministries was made clear in this experience.[85] The Farmers' Extension Service remained a major issue of stage II project implementation.

---

[82] See Benoir and Harrison 1976; Yeniceri 1980
[83] Average cotton yields, for example, increased from 1,700 kg/ha during the first year to over 3,000 kg/ha in 1969.
[84] The CEY is a World Bank supported extension programme to reorganize the extension service all over Turkey.
[85] The World Bank 1978, iii

3.2.2.3
Improving the Management: Irrigation Cooperatives and Irrigation Associations

The project was negotiated between the IBRD and the Turkish Government without the commitment to any special form of farmer participation for operation and maintenance. At the end of the stage I project Irrigation Cooperatives and Cooperative Associations were initiated, and they should contribute to an increased rate of irrigated land, to the improvement of services to the farmers, and to a higher collection rate of water charges.[86] The World Bank team underlined that „if more attention would have been paid to the organization of irrigator groups, recurrent costs would have decreased, and higher recovery could have been achieved".[87] Since 1967 Irrigation Cooperatives and Associations were established at on-village level, but they remained little more than administrative shells. In the Tarsus Plain, there were 12 active Irrigation Cooperatives, and 25 in the Yüregir Plain, with the Hacibali remaining as the only Cooperative Association. The basic unit of the farmers' organizations were the Irrigation Cooperatives, who were organized to receive water from one point of delivery, or grouped points of delivery. The Irrigation Cooperatives were incorporated in Cooperative Associations, their geographic boundaries being established to include irrigation cooperatives which had common problems. The responsibilites of the Irrigation Cooperatives comprised of planning and implementation of the water supply; encouraging that water was fairly distributed; the maintenance and repair of the water distribution system from the point of delivery; preserving the drainage system within their boundaries; the collection of water charges and enforcement of sanctions, etc. The Cooperative Associations would improve the service to the users by dealing directly with the cooperatives and the DSI. As a result of the establishment of Irrigation Cooperatives and Irrigation Associations, the rate of irrigated land increased in some areas, but until their reorganization in 1981, disputes continued over the collection of water charges, and on their abuse.

3.2.3
The Stage II Project and its Shortcomings

The stage II project was expected to provide irrigation and drainage networks and on-farm development for 39,250 ha, and on-farm works for an additional 9,000 ha which were originally part of stage I.[88] On an appraisal visit of the IDA, the subareas 3TP and 2YP with 16,000 ha, originally designated to stage III, were transferred to stage II. In both of these subareas no irrigation and drainage infrastructure existed except for some irrigation ditches and a lined lateral in the area 2YP. The land was used in dryfarming, and, in some few tracts, irrigation was practiced using either groundwater, or by pumping water from the River Seyhan. The absence of a natural or artificial drainage, and a high groundwater level fluctuating

---

[86] See Consortium TAHAL-ECI-SUIS 1973
[87] Ibid., p.39
[88] The World Bank 1985

between zero and 3 meters below the land surface had led to salinization of the area 2YP lands. Small depressions and the widespread accumulation of salt in the soil was directly related to drainage deficiencies. These areas required subsurface and deep secondary and tertiary drains to serve as collectors for subsurface tile drains.[89] A concrete pipe factory and O&M service stations would be constructed, project personnel would receive training overseas, and a study of cost recovery would be undertaken. The Farmers' Extension Service continued. The DSI's and TOPRAKSU's related works are listed in the Boxes 3.5. and 3.6. below.

Box 3.5. Sixth Regional Directorate for State Hydraulic Works - Stage II Project

Construction and extension of the main delivery channels and secondary lined laterals;
the construction of main drains;
the improvement and extension of the main drains;
a diversion drain to relieve an overloaded one, and lateral drains.

Box 3.6. Third Regional Directorate for Soil and Water - Stage II Project

On-farm development works comprised of land levelling, the installation of subsurface and surface farm drains on 28,730 ha, and irrigation ditches. In some areas, soil improvement was also necessary.

The consultant recommended some additional investigations to be carried out before the commencement of the works in the subareas 3TP and 2YP, e.g.,

* control of groundwater level fluctuations and drainage requirements in the 2 YP area;
* salinity control of the third Yüregir main drain (YD3) to decide on the amount of dilution needed to save the aquatic life in the Akyatan lagoon, and on the proposed alternatives of the YD3 drainage canal; investigation of the effects of chemicals in the drainage water on fish life;
* trend of the sea drift at the site of the proposed fifth Yüregir main drain (YD5) outfall to the sea, and the design of protection works to be provided below the proposed outfall structure of YD5;
* mapping of the actual property boundaries to enable the design of an economical layout of the distribution system.[90]

On-farm development works would start at the end of stage II in order that the relevant work could sequentially follow the project construction. The anticipated date of completion for all activities was March 1973. The total project costs were estimated at US$ 63.0 million; the IBRD/IDA funding covered 38% of the total

---

[89] See Consortium TAHAL-ECI-SUIS 1968
[90] Consortium TAHAL-ECI-SUIS 1968, pp.6-7

costs, the balance of the costs - US$ 39.0 million - was met by the Turkish Government. A loan of 587-TU and credit of 143-TU were approved in February 1969 (each US $ 12.0 million), and the final disbursements were made in November 1979.[91]

3.2.3.1
Evaluation by the IBRD/IDA and the Turkish Authorities

Two IBRD/IDA reports are available, i.e., the completion report for the stage II project which was prepared together with the Turkish Government, and an impact evaluation report.[92]

Time and cost overrun
The engineering works on a total of 45,700 ha were constructed without major modifications, and on-farm works had been implemented on 46,150 ha. Originally planned for completion in 1973, the construction of irrigation and drainage facilities was not, however, completed until the end of 1977, and on-farm development works in 1980. The reason for the delay was the need to renegotiate all contracts in 1974 as a result of increased inflation, the addition of 5,900 ha, and, foremost, the farmers' reluctance to participate in on-farm development.[93] Although TOPRAKSU successfully concluded on-farm development works, and land levelling was carried out on all suitable lands, on-farm development was seriously delayed from the side of the large number of farmers. When on-farm development should have been undertaken, the farmers refused to give up their fields, because cotton prices and yields had increased sharply. Cotton support prices per kilogram increased from TL 3.75 in 1973 to TL 6.00, to TL 8.00 in 1974, making the farmers reluctant to allow the work to be done during the cotton growing season. Many farm-owner operators rented out part of their land, or rented supplementary land, depending on family labor availability, capital, and consumption needs. Some landless farm worker families leased small amounts of land, others worked under share-cropping arrangements. None of these farmers could afford to lose the income from one cropping season. In the absence of any compensation for the loss of production and incomes, the farmers objected to the work on their fields prior to the harvest. They also refused to give up their supplementary rented fields for one growing season for TOPRAKSU's land levelling and drainage work.[94] Therefore, in 1973, „when land levelling was expected to increase from 5,000 ha in preceeding years to 15,000 ha, actual implementation declined from about 10,500 ha in 1972 to 6,000 ha in 1973, and then further to 2,600 ha in 1974".[95] Another reason

---

[91] See The World Bank 1985
[92] The World Bank 1978, 1985
The former report was based on a draft report of the Turkish Government, and on findings of a IBRD/IDA and a consultant mission. No separate documents are available from the DSI.
[93] The World Bank 1985, p.2
[94] See The World Bank 1985, p.11
[95] The World Bank 1978, p.9

that caused time delays was the land fragmentation, which made it difficult to organize operations rationally. TOPRAKSU could not obtain the permission to work in large blocks in any given season, and the costs were thus increased by the scattering of the work.

Within a 15 year development period and an economic life of 50 years, the project was expected to yield a rate of return to the national economy of about 16%, which was only 14.2%. There was a cost overrun of 18% (instead of US$ 63.0 million the actual costs were US$ 74.0). Most of the increase is accounted for by the increase in the quantity of work and local prices.[96] The on-farm development component increased substantially, and was attributed by the Turkish government as being largely due to heavier than anticipated land levelling that had been required, and to the increase in the area that was put under tile drainage from 20,900 ha to 22,400 ha. The composition of land levelling works is shown in Table 3.2.

Table 3.2. Land levelling requirements in the stage II project area (in percent of the total area)[97]

	Appraisal estimate [ha]	Actual [ha]
Light levelling or planing	52	31
Medium levelling	37	52
Heavy levelling	11	17
Total area	48,250	46,250

Drainage deficiencies: design and maintenance

The large drainways, and the secondary and tertiary drains were regarded as well designed. However, it appeared that more cross-sections could have been provided in open drains for sediment which deposited between regular cleanings. The new construction of much of the drainage networks had not taken into account, for example, controlled drain inlets, handling or shaping of the soilbanks, and seeding of drain ditch banks to stabilize against erosion. The maintenance of the constructed drainage systems had been inadequate, resulting in some secondary and several of the tertiary and open collector drains being clogged with silt and vegetation. Farmers apparently did not recognize the need to avoid silt inwash from their fields into the drain system, and had tended not to maintain surface drainage, and had even ploughed over them until they had disappeared altogether. In some cases, they had effectively destroyed the drains.[98]

---

[96] "There was no provision made in the estimates for inflation, a serious omission since the burden on the Government was considerably higher than anticipated." (The World Bank 1978, p.13)

[97] Source: The World Bank 1978, p.13

[98] See Consortium TAHAL-ECI-SUIS 1971; The World Bank 1978, p.20; ii

Inadequate water distribution and irrigation practices
Farmers usually took water directly from the canalettes with siphons. This resulted in inequitable distribution of water among the users. Each delivery made through a locked turn-out should be regulated by a ditch-rider only, who did not, at this point in time, exist in all villages; later, the ditch-riders would monitor the farmers' actions but not control the deliveries. The irrigation system had been designed without measuring devices except for Parshall type flumes[99] that had sometimes been included at the heads of some of the laterals. Individual turnouts, or measuring devices, were not provided for individual farms.

The farmers irrigated only during daylight hours, although the system had, however, been designed for 24-hour-irrigation, and excess water passed through the channels. As more area became developed, the system was to be used for 24-hours. Users downstream pumped water from the drains for irrigation purposes, e.g., for rice cultivation. Leaching of those lower lands would become necessary.[100]

Yields and crop diversification
Dramatic increases in both cotton and wheat yields were evident, but could not be sustained beyond 1971, because „wheat yields were affected by climatic factors such as low winter rainfall and the hot humid summer in 1972-73, and cotton yields by the severe white fly infestation of 1974-75. This infestation was brought under control by 1976 and yields recovered, but there was another setback in 1978 (late sowing due to late spring rainfall; infestation with red spider and resurgence of the white fly coupled with the resistence it has developed to the pesticides used). It is also evident that the reduction in yield levels coincided with the slackening off of the extension effort and with the increasing poor maintenance of drainage canals by DSI and/or surface drainage and land levelling by farmers".[101]

The key to exploiting the full potential of the Lower Seyhan Plain is diversification, involving a shift from monoculture of cotton (periodically complemented with wheat) to an intensive cultivation of horticulture and fodder crops in rotation with cotton, which involved double cropping and mixed farming. The area under cotton had always been high. Its reduction, however, was caused by the decreasing profitability of cotton due to pest infestations, which led farmers to increase cereal cultivation. This development in cropping pattern could be explained by the following factors: the white fly infestation reduced the average net income from TL 15,290 to TL 1,660/ha. Farmers suffered a net loss and, not suprisingly, the cotton area reduced drastically. The land thus shifted out of cotton was largely put under wheat, a risk-minimizing solution since it was a well known crop. This was also encouraged by the doubling of its support price from TL 1,200 in 1973 to TL 2,050 in 1974, and to TL 2,400 in 1975.[102]

---

[99] Parshall flumes are measuring flumes used to measure the flow of water in open conduits.
[100] See Consortium TAHAL-ECI-SUIS 1971
[101] The World Bank 1978, p.17
[102] Ibid., p.19

Cost recovery

Recurrent cost recovery was a major issue throughout the project II implementation, and a source of disagreement and disputes between the IBRD and the Turkish Government. Water charges for the DSI's O&M costs were always set at levels inadequate to cover actual costs, and even these were not fully collected. The O&M costs in 1970, for example, were TL 400/ha, compared to the actual average charge of TL 130/ha. In 1971, a cost recovery study submitted by the consulting TAHAL-ECI-SUIS made the following recommendations (see Box 3.7.):

Box 3.7. Proposals for cost recovery in the beginning of the project[103]

> b) The farmers' ability to pay, based on an assessment of incremental income derived from the project, net of taxes, risk and management allowance was estimated at TL 750/ha in 1970 prices. Thus TL 350 should be charged to recover capital costs, and this figure should be reviewed every 5 years to adjust to inflation.
> c) Existing penalties for non-payment which comprised a 10% surcharge imposed at the first repayment date should be increased with additional 2% to be charged each month, beginning with the second month of delinquency. Water delivery should be discontinued in case of prolonged failure to pay.
> d) The existing system of different charges assessed for different crops should be replaced by a single assessment rate for the entire project area except for the area under rice. In addition, water charges should be levied on the entire project area equipped with irrigation rather than only on the area actually cropped.
> e) Farmers' irrigation cooperative should be formed throughout the project area with responsibility to distribute water, operate and maintain the distribution system and collect water charges from the members.

The DSI accepted the proposal including the raising of water charges and increasing penalties for late payments, but it did not agree to levy a uniform charge per hectare regardless of the crop grown, because it believed at that time that this would encourage farmers to grow only cotton. The DSI forwarded its comment on the report to the Ministry of Finance, but the Turkish Government never officially adopted any of these proposals largely on political grounds.[104] The World Bank representatives made repeated probes, but after 1974, the Government implemented its own nationwide study. It resulted not only in this project being rated as a 'problem project' but also in the World Bank's decision not to consider financing further projects in the Lower Seyhan Plain until a satisfactory system was instituted.[105] In 1978, the newly elected Government substantially raised the water charges as a part of its general economic programme; but „nevertheless, with the proposed levels of capital recovery charges and unless collection rates are improved and a yearly adjustment made for inflation, recovery of project costs is

---

[103] Source: Consortium TAHAL-ECI-SUIS 1971, p.27
[104] The World Bank 1978, iii
[105] Ibid., p.27-28

likely to be negligible, at less than 100% of annual O&M costs only".[106] In the Lower Seyhan Project, the new charges were five times the previous levels, and averaged TL 900/ha compared with the actual 1977 O&M costs of TL 620/ha (or TL 760/ha of the actually irrigated area). Charges exceeded the previous year's costs not to account for inflation, but because water charges were considered to be an instrument of taxation, and thus the Lower Seyhan Irrigation Project area, being one of the wealthier regions of the country, was expected to subsidize more disadvantaged areas. The recovery rate ranged between 28% (1973) and 11% (1976), and the collection rate averaged at 89% (1973) and 49% (1974), and declined further at the end of the 1970s. The penalty system at that time was regarded to provide little incentive for farmers to pay. Particularly in conditions of high rates of inflation, this system effectively became an easy source of subsidized credit. In the event of persistent delinquency, the DSI had the right to confiscate property, but this right was rarely exercised and, as noted above, potentially more effective measures such as cutting off water or imposing surcharges on a monthly basis had been rejected. Both the Government's reluctance to enforce collection and the farmers' attitude reflected the political aspect of cost recovery.[107]

Capital charges were not collected until after completion of the works. High effective subsidy amounted to almost the entire investment costs, and it was with this consideration, that the IBRD informed the Government in 1976 that it would not consider financing irrigation until further steps were taken to improve the system.[108] In 1978, a decree was passed authorizing collection in the stage II area from 1979. All beneficiaries were required to repay the nominal capital costs over 40 years with no interest. If compared with the real economic costs, the level of recovery was very low.[109] The Turkish law did not provide for the imposition of cost recovery charges for the TOPRAKSU's works, and although the TOPRAKSU presented a draft in 1978 to remove this regulation, it had not been ratified.

3.2.4
The Stage III Project

The World Bank decided not to finance stage III of the project as its funds were to be used to develop areas with greater potentials, and because of the previously mentioned disputes over the issue of cost recovery. The Turkish Government had proceeded to implement this stage using its own resources. In the early reports, the area of stage III covered the lower lying parts of the plain with its difficult drainage problems. The feasibility report for the stage III project designated 52,800 ha (gross 60,900 ha) for development.[110] The area originally included the subparts 3-TP, 5-TP, the southern part of 1-TP in the Tarsus Plain, and 2-YP, 7-YP and 8-YP in the Yüregir Plain. Most of these lands presented special problems, e.g., high groundwater tables, waterlogging by intrusion from the river, saline and saline-

---

[106] Ibid., p. iii
[107] Ibid., p.29
[108] Ibid., p.30
[109] Ibid., pp.29-30
[110] The feasibility study was included in the stage II project.

alkali soils, and areas too low for gravity drainage. Due to high investment costs incurred, the area of stage III was reduced, and the most problematic part of the plain was designated for the stage IV project. Stage III as it was then implemented comprised of 19,830 ha (gross) with the subareas 1, 2 and 3-TP.[111]

Between 1976 and 1986, the DSI constructed irrigation and drainage networks on 17,780 ha (net), and drained 60 ha of swamp. The TOPRAKSU installed on-farm drains on 11,576 ha between 1981 and 1987, and the land levelling for all three stages was completed in 1980. Although the project was smaller in size than the stage I and II projects, the implementation lasted ten years because, at the end of the 1970s, the investments in the Lower Seyhan Plain slowed down due to po-litical disruptions. Then political decisions to allocate the resources to other re-gions and projects caused financial constraints, and the last stage IV project with its 40,000 ha was postponed.[112] The FAO evaluation team stated that these deci-sions reflected „the past emphasis given in Turkey to the initiation of new schemes rather than completion of existing ones".[113]

3.2.5
The Ongoing Rehabilitation Project: Drainage and On-Farm Development Works in the Stage I, II and III Project Areas

Under the Agricultural Sector Adjustment Loan approved in June 1985 to support the implementation of the Turkish Government's policy and institutional reforms for the agricultural sector, a „Core Programme of Priority Drainage and On-Farm Development Works" was agreed between the World Bank and the Turkish Gov-ernment. The objectives of the Core Programme for drainage and on-farm devel-opment works are the provision of adequate drainage in the service area of the DSI projects, and the implementation of on-farm development works. These measures were intended to eliminate waterlogging and salinity problems, prevent the further deterioriation of soils and restore existing irrigation areas to full production. The rehabilitation works in the drainage network operated by the DSI became neces-sary because they had been designed and constructed with a too low density, and also because the drains were not effective due to improper maintenance. The open drains were partially or fully silted up and full of vegetation. Existing structures previously constructed by the TOPRAKSU/GDRS, especially pipe culverts,[114] which obstructed the remodelling, or desilting of the drainage canals were demol-ished; on-farm drains were not maintained, and thus silted up. The causes of heavy siltation are of particular importance: the vegetation on the side slopes of drainage canals has protective effects, but it was also removed, and the side slopes have become unstable; side slopes of drainage canals were steeply built otherwise the expropriation of additional land would have been required. Embankments were

---

[111] DSI, 6.Bölge Müdürlügü 1988a, p.83
[112] DSI, 6.Bölge Müdürlügü 1986/87, 2.6
[113] FAO 1987, p.5
    Other evaluation reports for the implementation of the stage III project are not avail-able.
[114] I.e., a culvert made up of pipes of any suitable material (see DVWK 1983, No.7334)

constructed by the farmers to secure water in drought periods. The farmers met their irrigation requirements from pumped drainage water which is stored behind it. Some structures, previously constructed, were destroyed by the farmers hampering with the proper drainage of surface water, and also increased siltation. For example, small ditches which were excavated on the upstream side of plots to collect and lead water coming from surface runoff, are generally ploughed by farmers, so that the inlet structures became clogged, and do not function. The farmers then evacuated surface drainage water by cutting spoilbank. This damaged canal side slopes and enormous quantities of silt discharged into the canals.[115]

Within the scope of the programme, the Turkish Government envisaged a 7-year plan to rehabilitate the drainage infrastructure on an irrigable area (schemes already in operation) of approximately 720,000 ha. The projects, within the provinces of Adana, Antalya, Konya and Izmir, were divided into subprojects, of which the first project selected for implementation of the drainage programme was the Adana I subproject, being a part of the Lower Seyhan Irrigation Project area. Total investment costs are estimated to be US$ 255 million (US$ 74 million for the DSI's related works, the rest for the GDRS's related activities). It was agreed that the World Bank would finance 50% of the project costs (Loan No. 2663-TU). In November 1986, the General Directorate for State Hydraulic Works and the General Directorate for Rural Services invited engineering consulting firms to submit technical and financial proposals based on terms of reference issued separately by both organizations. In September 1987, both contracts were awarded to a joint venture of four engineering consulting firms.[116] The DSI and GDRS were made responsible for the implementation of the Core Programme.

The first project, Adana I-Subproject, started in 1989 and was completed in 1991. The Adana I-Subproject was selected as the first drainage improvement project for the following reasons: no major adaptions were needed in the main drainage system; drainage improvements were required in a substantial area; the area's agricultural potential was high; the project implementation time would be relatively short due to the already existing infrastructure. The experience of the GDRS officers in subsurface drainage system construction was more than adequate, and the previous field investigations could provide a substantial amount of data. Finally, the Internal Rate of Return (IRR), based on the provisional economic assessment, would, at 20%, be high.[117] The investment cost for the main drainage systems in the Adana I-Subproject is US$ 365.45/ha, and the cost per hectare for the on-farm works US$ 350 (prices 1995). In the Adana I-Subproject area, drainage is the major constraint to optimal agricultural production because 50% of the previously constructed subsurface drainage system (1968 to 1987) on 66,000 ha did not function properly due to the lack of maintenance, and the drain spacing (100m) was too wide for adequate control of the groundwater table in the heavy clay soils. The area of the Adana I subproject for drainage and on-farm improvement works covered 28,546 ha (11,921 ha in the Yüregir and 16,625 in the Tarsus

---

[115] DSI, DAPTA et al. 1994, p.16
[116] DSI, DAPTA et al. (undated)
[117] DSI, DAPTA et al. 1994, p.3

Plain). The DSI and the GDRS implemented the listed works in Boxes 3.8. and 3.9. The DSI's works were to be constructed before those of the GDRS.

Box 3.8. Sixth Regional Directorate for State Hydraulic Works - Rehabilitation Project

Rehabilitation and remodelling of open disposal systems construction of new open disposal canals Construction and rehabilitation of service roads

Box 3.9. Third Regional Directorate for Soil and Water - Rehabilitation Project

Installation of open quaternary and closed subsurface drains Installation of collector pipes Construction and rehabilitation of farm roads Reclamation of saline and alkaline lands (leaching and application of chemical amendments, e.g., gypsum)

The Core Programme will continue in the Lower Seyhan Project area with the Seyhan II-Subproject comprising of 8,346 ha. It includes similar works to the Adana I-Subproject, and, additionally, a pumping station (near the Baharli Village) for the drainage system and the re-shaping of the main drain TD0. The location of the other subprojects has not yet been identified (Adana II/Seyhan II with 8,346 ha; Adana IV/Seyhan III with 15,900 ha).

The rehabilitation programme implemented by the DSI showed delays caused by the lack of timely supplied funds. The most important bottleneck was fuel supply for the machinery, and the delay of expropriation allocations. For the construction of the structures, funds were only available for the excavation with the DSI's own equipment and machinery, but not for contracting out construction, which would have increased the costs.[118]

3.2.6
The Stage IV Project in Progress[119]

The not yet commenced stage IV project area, which originally was part of the stage III project, comprises 40,657 gross irrigable ha, of which 38,017 ha lie in the lower part of the plain near to the Mediterranean Sea, and 2,640 ha (6-YP) are located east of the city of Adana and north of the highway E 5. The lower parts of the project area include the subunits 5-TP in the Tarsus Plain, and 7-YP-8-YP in the Yüregir Plain. The construction of the required infrastructure had been postponed until the end of the 1980s due to high investment costs. The aim of this project is to increase the irrigable lands, and to initiate and improve fishery in the three lagoons Tuz Gölü, Agyatan Gölü and Akyatan Gölü. The lower parts of the stage IV project area show a tendency towards high groundwater levels, saline and

---

[118] DSI, DAPTA et al. 1994, p.9
[119] 6. Bölge Müdürlügü 1984, 1990

alkaline soils, and most of the lands produce low yields; 2,870 ha are saline, 24,455 ha are saline-alkaline (> 0.6% NaCl; > 15% sodyum), and 4,675 ha are alkaline soils. Part of the area requires leaching (2,870 ha), and 29,130 ha need leaching and the implementation of soil improvement measures (gypsum). Special drainage networks are necessary because the soil shows a low permeability, and there is no natural outlet to the sea. Although this area is not regularly connected, irrigation is practiced by withdrawing water from the irrigation and the drainage channels. Cropping throughout the year requires drainage pumping stations to lower the groundwater level to a depth of 120 cm. The irrigation networks will contain two main classical canals, irrigation channels (canalettes and pipe systems), four pumping stations for irrigation in the north, drainage channels and nine drainage pumping stations for the lower parts only. In the low-lying parts of the stage IV project area, 1,499 ha need surface drains, and 36,518 ha deep drains, of which 32,418 ha need on-farm drainage networks. Improving and initiating fishery in the three lagoons requires freshwater feeding channels and the installation of outlets to the Sea. The investment period is initially for 6 years; the estimated rentability is 2.65, and the internal rate of return 15.15% for the southern part. In 1991, the construction of the network in the Tarsus Plain (subunit 5-TP) was completed, the installation in the northern part is on-going. The southern networks in the Yüregir Plain will not be installed until the rehabilitation projects for the stages I, II and III, and the construction of the main drainage systems are ready. It is estimated that the stage IV project will be completed in the year 2001.

4.
The Effects of the Subprojects on the Distribution of Salt Affected Areas and High Groundwater Levels

According to the annual reports on groundwater and salinity and to research studies on soil, groundwater levels and soil salinity has changed with the gradual completion of the engineering works. Annual reports on groundwater tables and the salinity content have been available since 1966,[120] and studies from 1954/55 and 1982/84 document the distribution of saline and alkaline soils.[121] The changes in the distribution of salt affected areas in 1954/55 and in 1982/84 are shown by Table 3.3. and Map 3.3..

---

[120] Annual reports document precipitation data, the irrigated area and the crop pattern, and they provide estimations on the overall water-use efficiency. Data on groundwater levels and their salinity content (EC_{gw}) in the months with the most intensive irrigation, and their distribution within the investigated area are shown. (DSI, Genel Müdürlügü, Isletme ve Bakim Dairesi Baskanligi 1987, pp.9-14) (see Annex 5)

[121] See Nunns 1954/55; Dinc et al. 1986, 1990
A joint research project of the Institutes of Soil Sciences of the Çukurova University and the University of Stuttgart/Hohenheim prepared a soil salinity map which shows the formation, distribution, and chemical properties of the saline and alkaline soils in the Lower Seyhan Plain after thirty years of irrigation.

Table 3.3. Salt affected areas in 1954/55 and 1982/84 (in hectares)[122]

EC$_e$-values	1954/55	1982/84
4,000-8,000 micromhos/cm	19,982	35,941
8,000-12,000 micromhos/cm	29,054	17,760
>12,000 micromhos/cm	56,603	7,197
Total	105,639	60,901

The salt affected soils with EC$_e$-values between 4,000 and 12,000 micromhos/cm, and higher, decreased from 105,639 ha in 1954/55 to 60,901 ha in 1982/84. Within a span of 30 years, the soluble salts have moved and concentrated towards the southern part of the delta. According to the present soil salinity map, the salt affected soils, whose texture is heavy clay, are located mainly in the southern part of the Yüregir Plain. There, the drainage conditions are progressively deteriorating, increasing salinization and leading, eventually, to sodification. A physiographic barrier generates a depression without an outlet to the Mediterranean Sea.[123] These areas identify almost exactly with the area of the stage IV project area; they suffer because of the not yet constructed main drains, the drainage pumping stations, and delayed soil improvement measures.

As previously mentioned, the soils in the plain have developed on relatively impervious, recent deposits of the rivers, and have textures dominated by clay, silt, sand and gravel. The thickness of this stratum varies from 10 to 20 meters. Beneath this layer, both shallow and deep water-bearing aquifers and aquitards are encountered. The flat topography together with the low permeability of the stratum slow down the transmission of recharged groundwater towards the Mediterranean Sea, where percolated water cannot find an escape. If there are no adequate artificial drainage systems, the recharge of groundwater from surface sources results in a simultaneous rise in the groundwater table. Maximum groundwater recharge is observed in the winter months when the winter rainfalls approach the seasonal average; the annual critical peak level of the water table occurs in January and February, sometimes lasting until April (see Annex 4, Fig. A.1.). With lower than average seasonal rainfall, the highest water levels emerge in July and August, when most of the irrigation water is applied (see Annex 4, Fig. A.3.). The lowest water table is found in November/December, if groundwater is not significantly recharged during autumn rains (see Annex 4: Fig. A.2.).

From 1970 to 1974, the annual rainfalls were far below average, and therefore groundwater recharge from rainfall was minimal. Within the same period, high water table situations were encountered during the peak irrigation season in June, July and August. The groundwater level, which averaged at a depth of between 0 and 1 meter below the ground surface, remained more or less unchanged between

---

[122] Source: Dinc et al. 1986, p.6
    EC-values are given for the soil saturation extract (EC$_e$).
[123] See Dinc et al. 1986

1966 and 1975, except during extremely dry years. The area with groundwater levels between 0 to 1m increased from 45% (34,530 ha) in 1975 to 57% (57,784 ha) in 1989, when the 48,600 ha of the stage II and about 20,000 ha of stage III project were supplied with irrigation water. It can be concluded that the artificial drainage systems, which had been installed by 1986, only conveyed a share of the irrigation waters away from the area. The intensive drainage networks did not significantly lower the groundwater table. Despite the fact that on-farm drains covered an area of 2,500 ha in 1966, 41,000 ha in 1975, and 114,000 ha in 1985, the percentage of areas where groundwater levels averaged at depths between 0 to 1 meter increased. The excess water from winter rainfalls is drained away before the irrigation season starts in April/May, and the high groundwater levels in winter/spring usually create no problem for farming because planting starts in late April. With heavy and long-lasting spring rains, the installed drainage system is not adequate to drain the excess water, foremost, because of the wide lateral distances of the tile drains and the reduced capacity of the main drains caused by poor maintenance.[124]

The salinity content of the groundwater near the soil surface within the peak irrigation season shows a positive improvement between 1966 and 1989 (see Annex 4, Fig. A.4.).[125] The area percentage of groundwater with EC_{gw}-values greater than 2,000 micromhos/cm dropped from 52% in 1966 to 32% in 1975, and then to 30% in 1989. The area percentage of groundwater with EC_{gw}-values between 0-1,000 micromhos/cm increased from 16.5 to 34.5% in 1975, and those between 1,000-2,000 micromhos/cm has increased, in absolute terms, from 22,200 to 30,884 ha. This positive improvement can be attributed to the constructed drainage systems. The leached salts have not accumulated in the lower soil horizons, as they had before, they have continously been transported out of the area. The groundwater salinity contents show seasonal variations: available data indicate that the quality of groundwater during the irrigation season is better than in other seasons.

---

[124]The spacing of the tile drains is 100 and 150 meters, instead of 50 and 70 meters. (see Tekinel et al. 1976b, 1985)
[125]The salinity content of the groundwater near the soil surface is expressed in EC_{gw}-values which serve as an indicator for salinization in the monitoring programme of the DSI.

Map 3.3. Salt affected areas in 1954/55 and in 1981/84 [126]

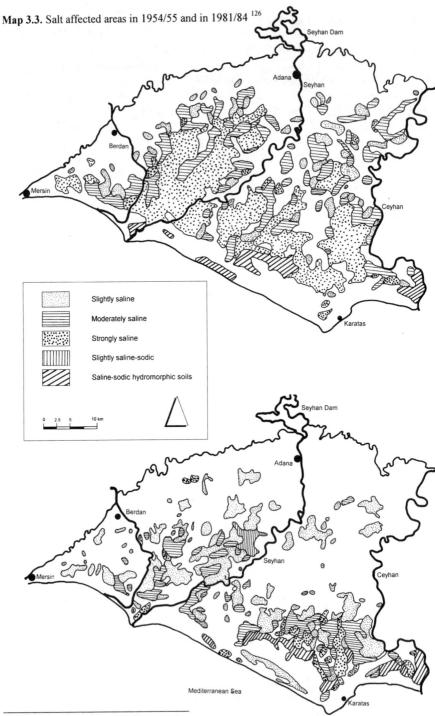

Slightly saline

Moderately saline

Strongly saline

Slightly saline-sodic

Saline-sodic hydromorphic soils

0   2.5   5   10 km

[126] Source: modified from Dinc et al. 1990, pp.150; 152

The salt accumulation in the plant root zone is greatly due to high evaporation. Autumn rainfall leaches the accumulated salts, and increases the salt concentration in the groundwater. When the soils become saturated during the rainy winter months, the quality of the groundwater deterioriates, and shows the highest salt concentrations. In spring, the highly saline groundwater is drained away through the drainage system, and consequently the quality of the groundwater improves.

When the first rehabilitation project was completed in 1991, positive effects on groundwater levels were immediately evident. Groundwater levels (0-1m) were analyzed during the period 1986/87 and 1991/92. According to the data, the area percentage with groundwater levels between 0-1m at the highest critical point reduced from 62% in 1986/87 to 47.5% in 1991/92. During the same period, the area percentage with groundwater levels of 0-1m at the lowest critical level decreased from 0.8% in 1986/87 to 0.4% in 1991/92 (see Annex 4, Table A.2.).[127] The area percentage with a groundwater level between 0-1m in the peak irrigation months decreased from 24.2% (1986/87) to 17.2% (1991/92), or in absolute terms, from 24,940 ha to 22,311 ha (see Annex 4, Table A.1.).[128]

Data for the Tarsus Plain, where the drainage and on-farm works started in 1988/89, show a decline in the area which had a groundwater level of between 0-1m from 4,000 ha to 2,810 ha in 1989/90.

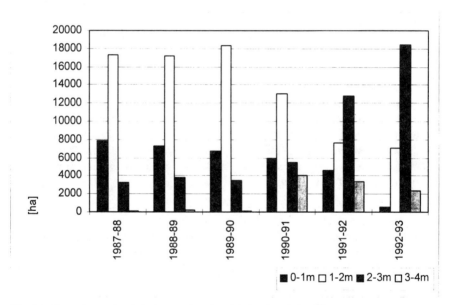

Fig. 3.6. Groundwater levels in the months of peak irrigation (in ha)[129]

---

[127] DSI, 6.Bölge Müdürlügü, ASO Subesi 1993; DSI, 6.Bölge Müdürlügü, ASO Subesi (Arca and Donma) 1994
[128] Ibid.
[129] Source: DSI, 6.Bölge Müdürlügü, ASO Subesi 1994 (Arca and Donma), Table 1

The percentage of area with a groundwater salinity content higher than EC_{gw} 5,000 mircomhos/cm decreased from 6.3% in 1986/87 to 3.7% in 1991/92. The final conclusions are that the percentage of lands with groundwater levels at depths between 0-0.5m reduced significantly; they were widely spread over the irrigation area in 1986/87. The levels 1-1.5m and 1.5-2.0m show a greater distribution in those parts where the DSI and GDRS had completed their works, and similar trends are observed with groundwater salinity contents of more than EC_{gw} 5,000 micromhos/cm. In areas where groundwater levels and saline soils are still high, there is, according to the DSI's studies, a great need for improvement in the on-farm drainage systems.

5.
Conclusions

High groundwater levels and salinity were subject to investigations in the planning stage. The feasibility studies for the subprojects identified the varying conditions together with the special drainage needs of the area, and defined the works to be executed by the state agencies responsible. Drainage requirements at on-project and on-farm level, and land levelling, ranged high on the list of priorities in the project outline for all subprojects. All engineering works within the area of the subprojects I, II and III were completed in 1986. During the implementation process, however, shortcomings and deficits in controlling groundwater and salinity occurred which are interpreted below (see Table 3.4. which provides an overview of the shortcomings at on-project and on-farm level).

1. **In the areas of the subprojects I and II, the installation of the main drainage networks gained high priority, and adequate resources had been provided. This trend was interrupted when the Turkish Government could not obtain external aid, and when it decided to direct the internal resources available towards other sectors and regions. Political decisions impeded the DSI's efforts to construct sufficient and effective drainage networks covering the whole project area.**

One outstanding difference to publicly financed large-scale irrigation projects in other countries is that drainage requirements were given major attention in the implementation process. The demand on increasing the irrigable area, or the expansion of irrigated agriculture, made it necessary to focus on groundwater and salinity control means in the planning and construction process, because high groundwater levels and saline/alkaline soils existed prior to the installation of the irrigation project, and it was well-acknowledged by the Turkish planning agencies and the international donors that the groundwater levels would rise with irrigation. The installation of drainage networks was viewed as the central control measure, and they were an essential in the planning and construction process of all subprojects. Sufficient financial resources, equipment and skilled manpower were provided for the Lower Seyhan Irrigation Project, and the construction of the main drainage networks at on-project level were carried out without major modifications by the Regional Directorate for State Hydraulic Works. Time overruns on the major works occurred due to contract procedures, and for the purchase of the DSI's equipment because of inadequate budget allocations. The drains at on-project level were designed and constructed with adequate capacities. More cross-sections in open drains would have been necessary for sediment deposits to be removed between the required cleanings. There was a lack of controlled drain inlets, of shaped soil banks, and of seeding on drain ditch banks for stabilization. The drain banks were steeply constructed to reduce costs for expropriations, and thus would have required additional stabilization. The main delivery channels have been concrete-lined, and earth-lined deliveries were substituted with canalettes, i.e., precast elevated concrete shells, which have served as secondary and tertiary channels. The canalettes have a lifespan of about 30 years during which they cost little to maintain. The higher investment costs have resulted in water savings, and have reduced the percolation of water. The main drainage networks within the areas of the subprojects were successfully installed, and the impact of the main engineering works was significantly positive as the areas affected by high ground-water levels and with saline/alkaline soils reduced within the service area (i.e., the stage I, II and III project).

The construction of the main drainage networks in the stage IV project area (40,000 ha) has been postponed, together with the construction of the delivery networks, with the result that, over time, high groundwater levels and saline/alkaline soils have concentrated in the southern areas. The area of the stage IV project was orginally part of stage III and required high investment costs for drainage pumping stations. However, with no external resources available, i.e., the World Bank decided not to support the further expansion of irrigation in the plain, the stage III project was reduced to a small area. It was financed from internal resources, and the construction works lasted almost 10 years.

The construction was postponed due to the Turkish government's decision to concentrate public investments in the energy sector, because Turkey suffered badly from the two oil crises during the 1970s. The exploitation of indigenous energy resources should limit the dependency of the national economy on imported oil. With the reorientation of the economic policies under Turgut Özal's government at the beginning of the 1980s, internal resources were primarily directed towards the South Eastern Anatolian Project (GAP) where water resources had not yet been exploited for energy generation, and where agricultural potentials had not been used to good advantage. Within the boundaries of the GAP, hydroelectrical power plants on the two rivers Euphrates and Tigris can provide 25% of Turkey's electricity demand, by producing 85% of its hydroelectrical energy, and 1.6 million ha of land can be irrigated. Another reason for investments in the south eastern provinces of Turkey was the increasing migration out of the region, and the expansion of the cities Istanbul, Ankara, Izmir and Adana; Istanbul, for example, has increased annually by approximately 400,000 people. Therefore, further investments in the Lower Seyhan Plain were postponed because the region was already well developed, and had, for a long time, received political support.

Table 3.4. Shortcomings in the implementation process of the subprojects

Subprojects	Financing investments	Implementation	Shortcomings at on-project level (DSI)	Shortcomings at on-farm level (TOPRAKSU)	Issues concerning O&M
Stage I 65,000 ha 1963-1968	Until 1966 (DSI) state-financed; no repayments from the beneficiaries (TOPRAKSU) farmer-financed through credits or own resources	(DSI) carried out with own forces; (TOPRAKSU) carried out with own forces on behalf of the farmers	without major modifications; difficulties in placing contracts and purchasing equipment because of insufficient budget allocations	great delays in on-farm works: finance investments not requested by farmers; insufficient credits due to poor allocations, and inconvenient credit terms	1968 Farmers Extension Service; 1968 Irrigation Cooperatives
			levels of drainage networks unadjusted; weak timely coordination of the works due to poor cooperation		
Stage II 48,600 ha 1969-1975/76	After 1966 (DSI) pre-financed through central budget; capital costs charged to the beneficiaries without interest; (TOPRAKSU) subsidized	(DSI) carried out with own forces and private contractors; (TOPRAKSU) carried out with own forces; rented out the machinery to the farmers at low prices	renegotiation of contracts due to inflation	great delays in on-farm works: farmers refused to give up fields for construction; no rational implementation; no investments in midway on-farm drains; credit terms remained inconvenient	Recurrent cost recovery issue

Table 3.4. (continued)

Subprojects	Financing investments	Implementation	Shortcomings at on-project level (DSI)	Shortcomings at on-farm level (TOPRAKSU)	Issues concerning O&M
Stage III 19,831 ha 1976-1986	(DSI) pre-financed through central budget; capital costs charged to the beneficiaries without interest; (TOPRAKSU) remain completely subsidized; proposal submitted for cost recovery not passed by parliament (1978)	(DSI) carried out with own forces and private contractors; (TOPRAKSU) carried out with own forces; after 1984 contracted out to the private sector	no information available	no information available	1981 village-based Water User Groups
Rehabilitation Project I 20,000 ha 1989-1991 Project II and III ongoing	(DSI) pre-financed through central budget; capital costs are charged to the beneficiaries without interest; (TOPRAKSU) changes in the investment strategy discussed	(DSI) carried out with own forces; (TOPRAKSU) private contractors and own forces	delays due to untimely supply of funds, and inadequate expropriation allocations	no information available	
Stage IV 40,657 ha TP 1991 completed YP not yet commenced	(DSI) pre-financed through central budget; capital costs charged to the beneficiaries without interest	(DSI) carried out with own forces	postponement of the construction of the delivery and drainage networks until the completion of the re-habilitation projects	installation of on-farm drains and soil improvement measures postponed until the completion of the rehabilitation projects and construction of main drainage structures	1994 Transfer of O&M responsibilities to Water User Associations

This region was no longer defined as a priority development area in the need of continous development support as it was until the 1970s. The investment decisions coincided with commitments set by the stand-by agreement signed between the Turkish Government and the International Monetary Fund. The economic memorandum of 1980, the basis for the negotiations of the first adjustment loan with the World Bank, supported projects to eliminate bottlenecks in the energy sector. Political decisions reflected the high opportunity costs of investments in the Lower Seyhan Plain, and directed public resources to other sectors and regions;[130] the effects within the project area were negative with high groundwater levels and salinity prevailing to date.

In the agricultural sector, running export-oriented projects, for which foreign financial resources could be secured, would be ended to optimally utilize their potentials.[131] Since the beginning of the 1980s, rehabilitation projects could be financed with external resources, and since the potentials of the Lower Seyhan Irrigation Project could only be fully exploited after the rehabilitation of the previously constructed facilities in the stage I, II and III areas, the Turkish Government applied for credits from the IBRD/IDA. Only when these conditions have been fully met can the implementation of the last remaining stage IV begin.

2. **The implementation of the on-farm development works was heavily delayed and insufficient, because the farmers could not afford investment costs, and access to credits was difficult. Even after changes in the investment strategy when sufficient resources had been publicly provided, the farmers were reluctant to cooperate.**

The most serious problems were related to the carrying out of the on-farm development works: land was inaccurately levelled; the installation of subsurface drains beneath the fields were inadequate; the initial tile drains had a spacing of 100 or 150 meters which was too wide. On-farm works, i.e. land levelling and the installation of on-farm drains, were seriously delayed during the implementation process of all subprojects. In the beginning, all on-farm works were to be financed by the farmers, and should have been executed by TOPRAKSU on behalf of the farmers, or by the farmers themselves. It was assumed that the farmers would request and finance land levelling and surface drainage themselves. The farmers, however, showed little response and interest in these investments, and no progress was made. This situation was complicated by the difficulties involved in obtaining credits, i.e., the conditions for obtaining credits were inconvenient, and the provisions from the central budget to the Agricultural Bank's and TOPRAKSU's funds were insufficient. As an early study of the DSI outlined, land levelling as a means for water saving did not make sense to the farmers, because water was charged at a flat rate, and water deliveries could not be measured. Or, to put it differently, it was not economic for them to substitute water for higher priced labour input.

---

[130] The major share of public investment funds was provided for the construction of the Atatürk Dam.

[131] See Wolff 1987, pp.112
The cutting of subsidies and price supports, and the taxation of agricultural incomes were further consequences of the economic memorandum.

In the stage II project, the cost overrun was largely due to heavier land levelling being required than had been estimated, and the increase in area to be put under tile drainage. Again, on-farm works were delayed: even with changes in financing after 1966, which had lifted the burden from the farmers, the farmers were reluctant to participate. In 1966, it was decided with the World Bank's advice that TOPRAKSU would plan and carry out the works free of charge. The farmers refused to give up land for levelling works, and for drain installation. Investments in farm drains midway between the state-constructed were still neglected by the farmers, for which there are many explanations: the on-farm works have no positive effects on the net incomes; the credit conditions were still inconvenient; the farmers were unaware of the benefit; net benefits were already high in cotton farming, or they feared government interference which would make on-farm development unattractive in the long term. Some farmers who had leased or rented out fields, or operated under share-cropping arrangements, could not afford the investments, and/or were not in a position to accept the loss of one year's income, as no compensations were foreseen for loss of production. Even large landowners did not invest. Whatever the individual reasons were, they have not received attention in planning and implementation.

The state agency TOPRAKSU/GDRS, who was responsible for the installation of the on-farm drainage systems and the execution of land levelling works in the areas of the subprojects, was confronted with great difficulties. Land fragmentation made rational operations difficult. TOPRAKSU, and later the private contractors, could not rationally organize operations; they could not work on large blocks at any one time, and TOPRAKSU had to deal with each farmer separately, which was time and money consuming. TOPRAKSU was legally not allowed, or able, to make contracts with the farmers before the construction works started. The farmers, who had not participated in planning, were reluctant to allow the works to be carried out, either because they already received high net incomes from cotton farming, or because they could not afford to lose income during the time it took for the works to be completed, and they therefore did not give up their land for construction works, or not at the time it was required. When the cultivation of wheat became a profitable alternative after 1971, the on-farm works could be executed more easily after the wheat harvest. After 1984, on-farm development works were no longer implemented by TOPRAKSU/GDRS's own personnel and equipment, but were contracted out to private firms which were expected to be more efficient. This shift in policy accelerated the works, but a major problem experienced in the course of implementing the on-farm drainage networks was still that contractors had often been unable to work because of standing crops, the attitudes of uncooperative farmers, and because compensation was not foreseen. The farmers' acceptance of drainage requirements was in general low because of higher expenses and their unawareness of benefits from drainage; on the contrary, they had not maintained them, and some farmers closed the installed drains for rice cultivation, and/or ploughed over them.

These problems occurred despite the fact that the main share of all public resources available for the on-farm development works in irrigated agriculture, including personnel, were concentrated in the Lower Seyhan Irrigation Project; the financing policy was completely changed by lifting the financial burden from the farmers, and that the private sector was involved which acclerated the works to be executed. The mentioned difficulties, therefore, cannot be attributed to financial constraints on public investments, or to the lack of manpower, etc., but to the top-down planning approach (without the farmers) and to the investment strategy which both have failed to provide incentives for the farmers to undertake the necessary works.

3. Coordination and cooperation between the two state agencies, and cooperation between the state agencies and the farmers was poor.

Although the project was jointly planned by the two state agencies, the construction of the main drainage networks and the on-farm drains were uncoordinated, resulting in unadjusted levels of on-project and on-farm surface drains; open drains and drain collectors were of insufficient depth for the tile drains. These uncoordinated efforts were time and cost-consuming. In Turkey, cooperation and coordination has always been a problem due to the highly centralized bureaucratic structure. After 1989, in the beginning of the rehabilitation project, coordination between the two state agencies was organized, by political decision, with the establishment of a joint committee.

The heavy delays in the execution of the on-farm works have shown the weakness of the top-down approach, and the inadequacy of both investment strategies. A national committee, recently formed with representatives from all concerned state agencies, in particular the DSI and TOPRAKSU/GDRS, demanded a new investment strategy where the farmers' requests would be given investment priority; not each project, that is technically, economically and politically approved, should be realized, and investments in undesired locations would not be made; the farmers would be organized in advance for operation.[132] The legal transfer of the completed facilities to farmers' organization is perceived to be the most appropriate and important state policy. The committee's proposal centres on active farmer participation, covering the whole process from project identification to the operation and maintenance of the newly implemented on-farm development works. The key assumption is that farmers who participate in actual decision-making, and who share in the costs, will also have a positive attitude and feel responsible for the maintenance and management of the on-farm infrastructures. The TOPRAKSU/GDRS' experiences, gained from the establishment of Water User Cooperatives, have shown the possibilities of successful farmer involvement to secure the sustainability of the projects. Instead of cost recovery, where the installations remain the property of the government together with the associated responsibilities, cost-sharing could guarantee that the farmers fully own the on-farm infrastructures. It would be clearly stated which part of the on-farm development investments is subsidized by the state, and which part is to be paid by the beneficiaries, including realistic interest rates. The farmers should commit themselves to sharing a substantial part of the on-farm development costs in real terms before any government support is given. The requirement of a cash investment by the

---

[132] See Sayin (ulusal calisma grubu) 1993; DSI, DAPTA et al. 1994; Kiratlioglu 1994

farmers should be abolished, or reduced, and the farmers should be offered a 100% financing package for the works through bank credits and TOPRAKSU/GDRS grants. The farmers' share should be financed through the Agricultural Bank, with the Bank taking the responsibility for payments to farmers and the recovery of the farmers' share. Priority should be given to organized farmers, as this would add to efficiency in terms of time spent by the TOPRAKSU/GDRS' staff on technical assistance, and the administration of credits and repayments.[133]

4. **The project was negotiated without making any commitments on the management of the scheme. The introduction of complementary means corrected the approach which centred only on engineering works.**

The project was negotiated between the international agencies and the Turkish Government without making any commitment on the management of the scheme. Both state agencies concentrated their efforts towards the installation of the engineering works. During the stage I and stage II projects, after operation had started, adequate water delivery attracted attention because no control could be exercised over the farmers who applied excess water, which resulted in inequitable water deliveries downstream. In addition, irrigation was only practiced during daylight hours, generating artificial water shortages downstream. Within the stage I project, the formerly earth-lined delivery channels were lined with concrete, and the installation of concrete canalettes solved the problem caused by some of the delivery networks being below field level, and of water having to be pumped up to the fields. This also reduced seepage from channels. Water deliveries were insufficient because the tertiaries had been overextended, and therefore rights-of-way problems emerged. No measuring devices, or turnouts from the secondary to the tertiary systems, and at the field outlets were installed; water has been withdrawn with siphons. As water deliveries were uncontrolled, unmeasured, and often insufficient, the establishment of Irrigation Cooperatives in 1968 should have overcome these deficits, but they had little effect, and the farmers were reluctant to join them. With the establishment of Irrigation Cooperatives, however, ditch-riders started to monitor water withdrawal. The Irrigation Cooperatives should also have played an important role in collecting the water charges from the beneficiaries, but did not succeed, and irreguliarities occurred in the administration of the water charges. Irrigation Cooperatives were successfully reorganized in 1981 as Water User Groups.

When the farmers had started irrigating in the stage I project area, a farmers' extension service was introduced because the farmers were inexperienced in irrigation practices. The extension service showed positive results, e.g., the farmers stopped flooding, and carried out irrigation by means of furrows in cotton farming. This extension service was later abolished; it did not receive support from the ministries because, as a special unit, it was not part of any national programme, and the government did not see any justification for continuing to allocate scarce

---

[133] It cannot yet be judged whether the new planning and investment approach, proposed by the state agencies, will be accepted by the government and passed by parliament. No attempt has been made, as far as I know, to implement it.

staff and other resources to an area which already received priority. No particular state agency was responsible, and had, thus, interest in supporting it. Farmers in some villages continued to employ the foremen until a new extension service was established.[134]

5. Legal regulations on recurrent cost recovery aiming to ensure adequate O&M budgets had little effects.

According to the project appraisal and the DSI's establishment law, the water charges should be assessed to fully recover recurrent costs. It was envisaged that the DSI would be responsible for the collection of adequate water charges, and that the beneficiaries' adequate contribution to O&M costs would ensure good performance. This would have rendered the DSI less dependent on the government's budget allocations by improving its own financial resource basis.[135] The DSI and the Irrigation Cooperatives were to play a crucial role: the DSI would assess the water charges, and the water assessment budget would be presented and approved at annual meetings of the Cooperative Associations, or the Irrigation Cooperatives. It was recommended that the Cooperatives would deny water delivery to members with delinquent payments. A fine of 10% of the bill was envisaged and an additional 2% of the bill would be added for each month, beginning with the second month of delinquency. Land with delinquencies of five years could be sold through sealed bids to the highest bidder. The Irrigation Cooperatives would be responsible for the collection of water charges; they would transmit assessments and penalty receipts to the DSI, and 20 % of the collected charges would serve to meet the costs of the Cooperatives.[136]

Some parts of these recommendations have never been accepted by the Turkish government, and the DSI has remained dependent on central budget allocations which form the ceiling for its O&M activities, while others could not be implemented because the establishment of irrigation cooperatives faced enormous difficulties.[137]

---

[134] This shows that the farmers were willing to pay for receiving services from which they could benefit.

[135] See IBRD/IDA 1963; The World Bank 1985

[136] See Consortium TAHAL-ECI-SUIS 1971, pp.3-7

[137] The recurrent cost recovery issue and O&M budget allocations are discussed in the following chapter in detail.

Chapter Four
Contributions to High Groundwater Levels and Salinity Caused by the Operation and Maintenance of the Lower Seyhan Irrigation Project

The central issue of this chapter is to identify the contributions to high groundwater levels, waterlogging and salinization by the operation and maintenance of the Lower Seyhan Irrigation Project. There, two management settings can be studied where the state agency and farmer groups share in O&M responsibilities: in 1981 village-based Water User Groups (WUG) with limited O&M responsibilities were established, and in 1994, Water User Associations (WUA) were initiated with a wider scope of autonomy. The working hypothesis is that in large-scale public irrigation systems institutional constraints bias towards the effective control of high groundwater levels and salinity if the state management unit is embedded in a highly centralized decision-making structure where the state agency is dependant on centrally allocated O&M budgets, and where the O&M budget allocations have nothing to do with the quality of the services provided, or with local O&M and salinity control requirements. If poor control mechanisms on performance, and if no accountability linkages are in place, public officials have no incentives for performing their duties well. The water charges, usually, do not instigate water-use efficiency, because they show no relation to the amount of water applied, and the officially introduced water allocation rules are poorly enforced although their enforcement could limit water distribution problems and could restrict excess water inputs. It is, furthermore, assumed that if irrigator groups participate in O&M activities, operation and maintenance will improve. The potential of the two public-farmer settings for controlling high groundwater levels and salinization is evaluated.

1.
The First Public-Farmer Setting: Joint Operation and Maintenance by the Regional Directorate for State Hydraulic Works and the Water User Groups

The first public-farmer setting has a highly centralized structure which can be described as follows: the state irrigation agency with its Lower Seyhan Branch is the management unit that depends completely on central budget allocations for

O&M. On the policy level, the State Planning Organization (SPO) and the Ministry of Finance (MoF) assess the O&M budget proposed by the General Directorate for State Hydraulic Works, and the legislature finally decides on the annual O&M budget allocations. The water charges which should legally recover the O&M costs from the project's beneficiaries, are assessed by a commission with representatives from the Ministry of Finance, the Ministry of Public Works, and the Ministry of Agriculture and Rural Services, with the final decisions being made by the government. The DSI is responsible for the planning and execution of the operation activities, and for the programming and execution of maintenance, in which it cooperates with Water User Groups. They both provide services for the 20,000 private farm units.

1.1
Areas of Responsibility of the DSI and the Water User Groups

The responsibilities of the Regional Directorate with its local branch for the Lower Seyhan Irrigation Project comprises operation; maintenance of the technical infrastructure facilities; monitoring programmes for groundwater and salinity, and the implementation of related control means.[1] At the upper management level, one deputy director represents O&M tasks within the Sixth Regional Directorate for State Hydraulic Works. One division is responsible for the operation and maintenance of all irrigation projects within the territorial boundaries of the Regional Directorate. The O&M division supervises seven local branches which are in charge of the operation and maintenance works, the groundwater observation programme and its implementation, and data collection, e.g., on cropping patterns. One branch operates the service area of the Lower Seyhan Irrigation Project, comprising 133,431 ha, with operational field units in each of the two water districts, the Tarsus and the Yüregir Plains.[2] The two main operation units are further divided into subunits, with engineers and technicians executing water distribution, maintenance, and repair works together with the village-based Water User Groups. The field operation unit of the Tarsus Plain water district is located in the village Yenice, with four subunits, and serves 69,031 ha. The operation unit of the Yüregir Plain water district, with its office in Dogankent, has three field units serving an area of 64,400 ha (see Fig. 4.1).

---

[1] See Annex 5: The DSI's responsibilities for groundwater and salinity control, observation guidelines and annual reports

[2] See Annex 1: Data on the service area and the irrigation and drainage networks

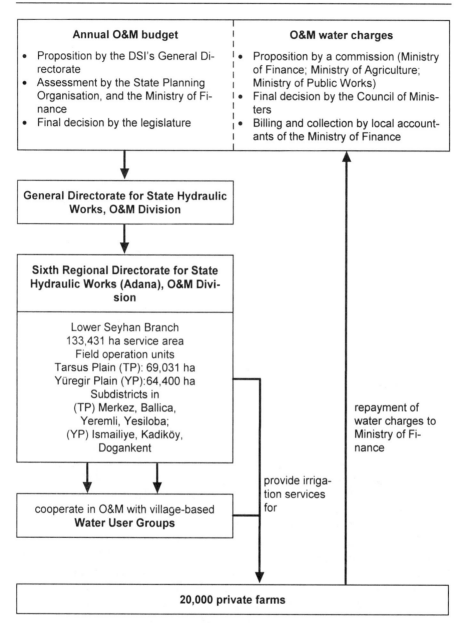

Fig. 4.1. Parties concerned with operation and maintenance (O&M) in the first public-farmer setting

Table 4.1. Areas of responsibility of the state agency and the Water User Groups

	DSI, 0&M Division	Water User Groups
Water supply planning Operation	pre-seasonal, current; from main to secondary networks	collection of the farmers' preseasonal and current water demand papers; from secondaries to the farm outlets; control of the schedule and rights-of-way
Conflict settlement	between WUGs	among irrigators of one village
Maintenance and repair	main deliveries; secondaries and tertiaries (mechanical works)	secondaries and tertiaries (manual works only) repair only if contracted
	drainage networks	not involved; but should prevent misuse, grazing on the banks, plantation of trees; manual weed control
	control of WUG maintenance activities	field inspections by DSI and WUG headmen
Observation programme on groundwater and salinity	DSI's staff only	

Village-based Water User Groups are responsible for a specified area of operation activities, i.e., the secondary and tertiary level until the field outlets, and for maintenance works in the secondary and tertiary delivery channels. They have to ensure the water delivery to the plots; their ditch-riders monitor whether rights-of-way are respected, and whether the water supply schedule is respected and followed by farmers at the field outlets. The WUGs collect the farmers' preseasonal and actual water demand papers, and participate in manual maintenance works, e.g., manual cleaning from sediment, minor repair works, and the eradication of weed along the channels. They are, however, not responsible for the drainage systems, but they should dissuade the farmers from using the drainage networks as dumps. Their headmen have to solve disputes amongst the irrigators if rights-of-way are not respected, or if the supply schedule is opposed. Depending on the seriousness of the rule violation, conflicts are to be settled within a WUG, or with the assistance of the DSI's personnel, or before the local court. The DSI's responsibility is actually limited to the main systems' boundaries, and a WUG should try as much as possible to solve disputes itself. Because the Water User Groups operate within one village, the DSI's staff continues to play an important role in coordination and conflict resolution if more than one village is supplied with irrigation water from one channel.

1.2
The Regional Directorate for State Hydraulic Works

1.2.1
Staff and Equipment

The Sixth Regional Directorate for State Hydraulic Works employs skilled and experienced personnel who are not as frequently, or to such a great extent, displaced with political changes, which is a common feature in other Turkish ministries. The DSI has initiated training programmes for its engineers in Turkey as well as abroad, and because salaries are based on education received, and on seniority, the engineers find the training prospects very attractive. It has always been difficult to attract highly qualified personnel to work in the underdeveloped regions of Turkey, in the south east, for example, and particular incentives have been set up. The Çukurova region, however, with its cities of Adana, Mersin and Tarsus, is one of the main attraction centres of Turkey, after Istanbul, Ankara and Izmir, with highly developed industries, commerce, infrastructure and transport facilities, and a high standard of living. Compared with the private sector, however, public sector salaries for engineers are low. Beginning with the multiparty period in the 1950s, the bureaucracy was subject to severe pressures which eroded its former high quality. The salaries and emoluments in the public sector underwent a serious decline as a result of rapid inflation, and increases in salaries in the 1970s were not really sufficient to make them an important factor in attracting the best talent to the public sector. Even today, the salaries for public officials are extremely low when compared with the private sector, and with the salaries of publicly employed labourers. Despite these financial disadvantages, the public sector provides some privileges for its employees: the 5-day working week, regular working hours and holidays, favourable credit terms, holiday accomodation available at special discount prices, cheap housing, etc. For irrigation engineers, however, unlike construction engineers, the public sector remains a favourable employment opportunity. Among the DSI's staff, there is a strong organizational sense of belonging to the DSI's establishment, and strong professional ethics that the works to be done should be state-of-the-art. Well skilled and experienced persons became the directors of local O&M branches, and good performance of irrigation projects, or even of subdistricts, is rewarded, at least for some officials, with advancements within the hierarchy.[3]

Although the performance does not solely depend on the numbers of staff employed and on the equipment available, sufficient staff and equipment remain necessary conditions for good performance. Whether the numbers of staff employed to control operation and maintenance are sufficient or not is an issue of controversy. The DSI claims a lack of qualified staff, but an FAO team emphasizes that the staff employed is adequate in number when compared with other large schemes. The

---

[3] These considerations are based on individual observations, and insights, made during many stays in Turkey, meetings with different authorities in several regions of Turkey, and especially through the assistance of the General and the Regional Directorate for State Hydraulic Works.

O&M costs per hectare are quite high, and would be sufficient to employ adequate manpower (they are higher than the costs of other schemes).[4] Bos and Nugteren define 0.35 personnel per 100 ha as an indicator. In the Lower Seyhan Irrigation Project, 0.32 personnel (without temporary employed labourers) and, together with temporarily employed labourers, 0.55 personnel are available for 100 ha. [5] (see Table 4.2.)

One delivery channels in the project area which is 32m/ha compared to an average of 14m/ha. The DSI employs 0.10 employees/km. The indicator is 0.05/km for the most highly-mechanized and high-waged countries, and 0.5/km for low-mechanized and low-waged countries. The DSI's staffing ranges in the same order as some Italian, Canadian and Californian schemes,[6] and compared with other international projects, the size of the DSI's establishment does not constitute an exaggerated size in relation to its tasks.

Table 4.2. Number of staff for O&M per year and service area[7]

Year	Staff[a]	Service Area	Staff per 100 ha
1976	404	98,547	0.40
1977	479	104,102	0.46
1978	468	110,480	0.42
1979	522	110,480	0.47
1980	507	103,000	0.49
1981	398	103,000	0.38
1982	414	103,000	0.40
1983	407	115,000	0.35
1984	409	119,000	0.34
1985	417	125,300	0.33
1986	386	132,300	0.29
1987	n.a.[b]	n.a.[b]	n.a.[b]
1988	432	131,700	0.32

[a]   The numbers of staff do not include temporary employed labourers.
[b]   no data available

The main problem concerning the DSI's activities is labour productivity, and the incompatibility between the working hours and the operation time. The technical

[4] See DSI, 6.Bölge Müdürlügü 1986/87; FAO 1987; 1988
[5] Bos and Nugteren 1974, Table 5
In 1988, the Lower Seyhan Branch employed 744 persons, (i.e., 39 technicians, 7 non-technical personnel, 386 permanent workers and 312 temporary workers). The 39 technicians are 8 civil engineers, 6 mechanical engineers, 18 agricultural engineers, 1 chemist, 3 bricklayers, 3 machine fitters. (see DSI, Sixth Regional Directorate 1988)
[6] See The World Bank 1992, p.36;86
[7] Source: FAO 1988, pp.11;31; DSI, Genel Müdürlügü, Isletme ve Bakim Dairesi Baskanligi 1989, p.156

personnel, usually the water distribution technicians, officially start at 8.00 am in the morning; the working day ends, officially, at 17.00 pm. But they are not in the field before 9.00 am, normally later. Then, between 12.00 and 13.00 is the midday break, which usually comes to an end after 13.00. Nobody is to be found in the field later than 16.30. Before the Ciller government came to power overtime was paid, which was very attractive for the personnel. No money is available now, however, to harmonize the working hours to fit with the operation times. Under the present conditions, the DSI does not see any possibility for change, and for improving labour productivity because of the strength and influence of the trade unions. Increasing the number of employed personnel, or providing extra-salaries, is also impossible due to budget constraints. There is no lay off of employed personnel; but the DSI's operational staff, however, has not been increased for six years, despite an annually expanding service area of between 70,000 and 80,000 ha (nationwide). Staff can be exchanged or transferred between the state agencies, and between the Regional Directorates of the DSI, but new personnel cannot be hired due to a halt in recruitment. Although part of the programme of the Turkish Government under Tansu Ciller is to improve the efficiency of the public sector, the coalition with the social-democratic party makes radical changes in the personnel policy impossible.

The DSI carries out all routine annual maintenance and repair (i.e., sediment removal and weed control, production of canalettes and their replacement, road grading) with its own equipment and machinery. This practice dates back to the beginning of the project where no private contractors could procure and operate the specialized equipment needed. Now, a large stock of equipment and vehicles with large repair facilities is still maintained. The government-owned equipment usually works at a fraction of the productivity of machines operated by the private sector. In an evaluation report, the World Bank team interprets this strategy as a result of the government considering the large amount of existing DSI equipment and workshop establishments as sunk costs, and „the position of the personnel employed to operate and maintain all this equipment, even when their operating hours and productivity is a fraction of that in the private sector, as inviolate".[8] The present problem is that if this strategy would be replaced with contractor-executed routine maintenance, while maintaining a much smaller stock of equipment for emergencies, higher cash allocations would be necessary under constantly decreasing budgets. The factual results are that during the 1980s, only approximately 50% of the total maintenance programmes could be executed.[9]

The inadequacy of machinery for the drainage systems' maintenance is an undisputed issue.[10] Due to the shortage of machinery, the DSI has accumulated a large backlog of drainage maintenance work. The machinery and equipment, bought about 20, 25 years ago to serve approximately 70,000 ha, is now quite

---

[8] The World Bank 1992, p.86
[9] See DSI, 6.Bölge Müdürlügü 1986/87; DSI, 6.Bölge Müdürlügü 1989a
     For example in 1988, 1,103 km channels should have been cleaned and about 7.6 million cubic meters of sediment should have been removed, but actually it was 590 km and 3.5 million cubic meters.
[10] See The World Bank 1978; DSI, 6.Bölge Müdürlügü 1986/87; FAO 1987; 1988

aged. Since then, the machinery stock has not been renewed; in 1983/84, only one excavator, one bulldozer and one grader were bought. Now, however, most of the machines are defect, and the numbers reduced to the level of 1984. Although the service area comprises, meanwhile, 132,000 ha, the machinery and equipment has remained at a level to serve only 70,000 ha.[11]

1.2.2
Financing Public Irrigation Services

The present funding procedure for operation and maintenance consists of two separate and unrelated issues: firstly, the estimation of the regional O&M require- ments and the decisions on the DSI's O&M budget, and secondly, the assessment of the water charges, and their billing and collection from the farmers.[12] (see Fig. 4.1.)

1.2.2.1
Regional O&M Requirements and Decisions on O&M Budget Allocations

The Regional Directorates estimate their O&M requirements on the basis of annual inspection reports which particularly specify the requirements for the cleaning of the delivery and drainage channels, the repair of canals and canalettes, excavation works, road maintenance, and the like.[13] The requirements are fixed per scheme in physical units, and, on the basis of the DSI's unit rates, published each year and after being corrected for expected inflation during the coming year, and the avail- able equipment, they are converted into costs. The regional inspection reports with the estimated O&M requirements are submitted to the General Directorate at the end of the year. The regional estimates are reviewed and corrected by the General Directorate, notably the Budget Department and the O&M Department.[14] The adjustments are made for the DSI's own personnel, equipment, and machinery, and for the budget's cash parts in the light of information obtained from the State Planning Organization and the Ministry of Finance about the likely size of the new budget, making a judgement about what maintenance works can be deferred. The calculation of the estimated regional O&M requirements is completed by July in order to submit the overall DSI budget proposal, which is then reviewed by the State Planning Organization and by the Ministry of Finance. Unspecified correc- tions are made by parliamentary decisions, for example, deciding that all alloca- tions should be cut by 2%. The parliament usually follows the SPO's proposal after it has been examined by the Budget Plan Commission.

---

[11] See 1977-1986 Yillari Makina Parki, in: DSI, 6.Bölge Müdürlügü 1988a, pp.34-35
[12] The presentation of the funding procedure benefits from a World Bank study (1992), DSI data and interviews with the DSI's staff from the Department of Operation and Maintenance of the General Directorate.
[13] DSI, Genel Müdürlügü 1995, pp.38
[14] The World Bank states that "the DSI calculates its expected O&M funding require- ments in a professionally sound way". (1992, p.8)

The DSI's annual budget mainly consists of construction expenditures (including beside others, maintenance expenditures), wages and salaries, and purchases (e.g., energy, fuel, machinery, spare parts).[15] Because the DSI carries out the operation works with its own forces, i.e., with its own personnel and its own stock of equipment, the costs are divided between the DSI's costs for its own forces and the required cash components. The maintenance cost for the DSI's own forces comprises equipment and staff expenditures; the earmarked cash components comprise of repair works, using purchased materials, spare parts, hired labour, and payments to the contractors. The DSI's requests are reviewed by the State Planning Organization and the Ministry of Finance, in particular the cash requirements for maintenance/repair works which are, due to their flexible nature, subject to cuts by the SPO, while the costs for operation, consisting mainly of salaries of permanently employed personnel, are usually added. The Ministry of Finance makes only minor changes since a large portion of the part under its review consists of wages, salaries and pensions, and then fuel and spare parts. Since the beginning of the 1990s the maintenance share of the O&M costs declined while the operation component within O&M, consisting mostly of salaries, increased. (see Table 4.3.) In 1992, for example, the DSI's request for maintenance was reduced by 66% by the SPO, and allocations for salaries were increased by 4.8% by the Ministry of Finance.[16]

Table 4.3. Maintenance and personnel costs as a percentage of O&M costs[17]

Year	0&M costs [TL/ha]	Maintenance costs [%]	Personnel costs [%]
1983	177,863	62	n.a.[a]
1984	184,275	54	18
1985	195,293	47	12
1986	187,846	47	16
1987	168,944	54	16
1988	140,713	53	13
1989	163,876	32	13
1990	161,760	33	29
1991	n.a.[a]	29	41
1992	n.a.[a]	n.a.[a]	50

[a]   no data available

Table 4.4. shows the reduction of the allocated cash component for maintenance and repair expenditures compared with the requests of the Regional Directorates from 1986 to 1994. The regional cash requests for maintenance show an increasing

---

[15] See The World Bank 1992, p.83
[16] Instead of TL 9.79 trillion, TL 3.33 trillion were allocated for maintenance, and for salaries TL 3.63 trillion instead of TL 3.80 trillion were approved. (Ibid., p.34)
[17] Source: The World Bank 1992, p.88

trend, while the allocated cash component for maintenance and repair, compared with the regions' requests, is substantially reduced with the result that, from 1990 to 1994, all maintenance and repair programmes could not be fully executed (i.e., less than 50%) due to budget shortages. The General Directorate of the DSI freely decides on the regional distribution of the maintenance cash component within the earmarked segments, and the reductions are distributed in a proportional way over the regions. The years 1993 and 1994 show irregular figures with a high percentage of allocated cash; a reason could be that some schemes were intended to be transferred to Water User Associations, and only those schemes in good shape could be transferred.[18]

Table 4.4. Regional maintenance requirements and allocations from 1986 to 1994 (TL million)[19]

Year	Estimated regional maintenance cost	Regional cash request as % of maintenance	Percent cash allocated
1986	16,62	43.7	70.3
1987	27,17	39.8	45.8
1988	37,31	37.6	42.5
1989	60,89	56.5	32.5
1990	63,04	60.2	44.2
1991	93,30	55.7	48.4
1992	120,00	75.0	31.0
1993	219,00	72.0	79.2
1994	423,50	32.0	111.2

1.2.2.2
The Assessment, Billing and Collection of the O&M Water Charges

According to the DSI's establishment Law No.6200, the irrigation water charges should be sufficient to cover the annual operation and maintenance costs. But this target has never been achieved, once because of the lower than required assessments of the water charges and their poor collection, and, above all, due to the funding procedure. The present funding procedure works ex-post facto. To cover O&M expenditures that took place in year 1, the assessment is calculated and approved in the spring of year 2; the farmers are informed of their assessment at the end of year 2, and the tariffs for year 1 are collected by April of year 3. Thus, current expenditures are 'financed' by the collections for irrigation services of two years ago.[20] The charges are determined by dividing the previous year's O&M expenses by the irrigated area. Tariffs are set for six different groups of areas, depending on whether the areas are economically 'underdeveloped', or involve

---

[18] See 2. in this Chapter
[19] Source: The World Bank 1992, p.35; DSI, Genel Müdürlügü, Isletme ve Bakim Dairesi Baskanligi 1995, p.5;7
[20] See The World Bank 1992, p.32

pumping, or only drainage, etc. Within each of these six groups, there are different tariffs for 29 crops based on differing water requirements, and on assumed differing profitabilities of the various crops. The stage of implementation and operation is considered as well.[21] To recover the expenditures for the drainage systems' maintenance, the water charges are calculated and assessed with respect to the size of irrigated land, but irrespective of the crop and the applied amount of water. Other users of the drainage systems, for example, the city of Adana and the industries, are not charged and do not have to contribute to the drainage systems' maintenance costs although they increase the maintenance requirements.

The tariffs are submitted by the Ministry of Public Works to an inter-ministerial commission, and presented for approval to the Council of Ministers. The commission is formed by representatives of the Ministry of Finance, the Ministry of Agriculture and Rural Affairs, and the Ministry of Public Works. Law No.6200 (Art.29) defines that the water charges are subject to reductions by the Council of Ministers („The Government may reduce the proposed rates.") which usually sets the water charges lower than the costs incurred to encourage the farmers to give up non-irrigated agriculture, and to practice irrigation. Therefore, in the early stages of the project, water charges were set at a very low level. Because the farmers were not able, or not willing to invest in irrigated agriculture, the low charges were to stimulate the shift from dryfarming to irrigation. They were not to impose an additional obstacle for the farmers as the input costs per hectare of irrigated cotton farming, for example, were already four times higher than in non-irrigated cotton farming, and the average per capita income of farmers was one-fifth of the average per capita income in other sectors.[22] Since 1982, the government has operated under fiscal constraints, and therefore, after 1984, the water charges were assessed at higher levels, which approached full recurrent cost recovery.

In an annual decree,[23] the tariffs are published two months before the irrigation season starts, at the latest in April, so that the farmers are well informed. The water charges are then billed for the individual farms according to a farmer's water demand paper. The charges relate to the size of irrigated land and to the crop pattern. The DSI's staff together with the ditch-riders of the Water User Groups annually measure the size of the irrigated plots as a basis to obtain the correct size for these calculations. The individual bills are published in the villages in October, and they are passed on to the beneficiaries, who then have a month to settle disputes, and make adjustments. The final billing is established at the end of December by the Regional Directorate's Accountants Office, which comes under the Ministry of Finance. Payments have to be made in two instalments, e.g., January/February and March/April. Fines for non-payers, when they are imposed, are set at minimal levels. Should the charges not be paid until the end of the year, a one-off penalty is

---

[21] The legislature distinguishes between three groups of farm enterprises: (Group 1) Farms where installations are still in the process of being built. (Group 2) Farms where installations have been completed and which have been in operation for between 1 and 10 years. (Group 3) Farms where installations have been completed and which have been in operation for 11 years. (see Metin 1971)

[22] Bülbül 1995, personal communication

[23] See Annex 6

set at 10% of a farmer's O&M charge.[24] Apart from the ex-post facto assessment procedure, the actual collection of the water charges, and the fines, has been a problem of major concern. The job of collecting the water charges and the fines is the responsibility of special collectors under the Ministry's accountant, attached to the regional DSI offices. The Accounts Office is understaffed, the fee collectors have transport problems, and thus the collection ratio usually ranges lower than 50% of the assessed charges (see Table 4.5.). Its personnel is poorly paid, and it is subject to widespread political interference to reduce or eliminate the payment of charges for entire villages or individual farmers. The collection of water charges for irrigation services rendered about two years ago in an environment with high inflation and the lack of sanctions for non-payers is, however, discouraging. Under the present institutional setting, the administrative costs for collecting the charges and the fines are higher than the money received.

Table 4.5. Assessed water charges as percent of O&M expenditures, and rate of collection[25]

Year	Assessment as percent 0&M cost	Rate of collection
1984	61	53.1
1985	90	49.2
1986	99	47.4
1987	72	42.2
1988	97	37.6
1989	104	37.6
1990	90	36.0
1991	90	32.4
1992	70	n.a.[a]

[a]   no data available

1.2.2.3
O&M Expenditures in the Project Area, and Their Relation to the Assessed and Collected Water Charges

In the 1980s, the main part of the O&M expenditures in the Lower Seyhan Irriga-tion Project referred to maintenance and repair works, with the highest share of 82.5% in 1984 and the lowest share of 69.9% in 1988 (the national average was 54% in 1984, and 53% in 1988). During this period, the operation cost component for the Lower Seyhan Project was lower than the Turkish average, but it has sharply increased during the last five years. Operation costs accounted for 67% (1990), 70% (1991), 71% (1992), 73% (1993), and 75% (1994) of the total O&M

---

[24] This regulation draws heavily from American irrigation legislation.
[25] Source: The World Bank 1992, p.80; DSI, Genel Müdürlügü, Isletme ve Bakim Dairesi Baskanligi 1995, p.11

costs. Personnel costs range from about 45 to 49% of total O&M costs.[26] In 1994, the personnel costs for the DSI-operated left bank water district reached almost 75.7% of the total costs, and the maintenance share was only 14.6%. The maintenance shares decreased from 33%, in 1990, to 25% of the total O&M costs in 1994 (see Table 4.6.).

Table 4.6. Itemized operation and maintenance expenditures between 1990 and 1994 (percentage)[27]

Items	1990	1991	1992	1993	1994
Total operation cost	67	70	71	73	75
Personnel	44	45	46	49	42
Energy/Fuel	11	14	14	12	21
Equipment	10	9	9	9	9
Other	2	2	2	3	3
Total maintenance cost	33	30	29	27	25
Total 0&M cost	100	100	100	100	100

Table 4.7. Cash component as a percentage of the maintenance expenditures from 1986 to 1995 (TL million)[28]

Year	Total maintenance cost	Cash component maintenance[a]	Cash percent of maintenance cost
1986	2,810	0,790	28.1
1987	3,429	0,716	20.8
1988	2,480	0,480	19.3
1989	3,255	0,892	27.4
1990	3,462	1,070	30.9
1991	6,000	1,600	26.6
1992	7,674	2,016	26.2
1993	8,850	6,100	68.9
1994	15,000	10,661	71.0
1995	14,157	1,684	11.8

[a] construction materials, expenditures for subcontractors

[26] DSI, Genel Müdürlügü, Isletme ve Bakim Dairesi Baskanligi 1995; DSI, Genel Müdürlügü 1995
[27] Source: Ibid.
[28] Source: DSI, Genel Müdürlügü, Isletme ve Bakim Dairesi Baskanligi 1995

The regional O&M requests were usually reduced by 30 to 40%. The cash component of the maintenance costs shows no clear trend (see Table 4.7.). The allocations for maintenance increased in 1993 and 1994, probably because the scheme could only be transferred to the Water User Associations after the execution of maintenance works. In 1995, maintenance responsibilities were taken over by Water User Associations, and the DSI's contributions have been limited, and thus the expenditures reduced.

The O&M expenditures are only partly assessed to the beneficiaries, as the following data indicate, but the assessments have increased substantially since 1984 (Table 4.8.).

Table 4.8. Assessed water charges as a percentage of O&M costs, and rates of collection in the project area between 1970 and 1988[29]

Year	O&M costs	Assessed charges	Assessed as % of O&M costs	Rate of collection
1970	290	130	44.8	
1971	430	150	34.8	
1972	360	170	47.2	65.0
1973[a]			43.0	89.0[c]
1974[a]			46.0	49.0[c]
1975[a]			37.0	67.0[c]
1976[a]			16.0	
1977	1,160	180	15.5	
1978	740	160	21.6	2.3
1979	1,500	810	54.0	2.2
1980	2,850	800	28.0	2.4
1981	6,770	760	11.2	4.2
1982	8,120	950	11.6	8.1
1983	11,050	1,390	12.5	22.1
1984	15,350	3,090	20.1 (61)[b]	27.9
1985	16,820	6,110	36.3 (90)[b]	29.7
1986				20.7
1987				22.6
1988				19.5

[a] The data for the years 1973 to 1976 are from The World Bank 1985, p.29.
[b] The figures in parantheses are the percentage of assessment related to O&M costs on a country-wide level. The difference between the numbers is unclear.
[c] Rate of collection in the stage II project area from 1973 to 1975.

The DSI's evaluation report for large-scale irrigation systems emphasizes that the DSI would have accepted higher rates, and that, in its view, this issue has not been adequately considered by the politicians, and high subsidies and inadequate budg-

---

[29] Source: FAO 1988, p.113; The World Bank 1985, p.29; 36; DSI, Genel Müdürlügü, Isletme ve Bakim Dairesi Baskanligi 1989, pp.208-209

ets for O&M works have remained the usual feature.[30] In 1979, for example, the newly elected government raised the charges substantially as a part of its general economic programme. The water charges were raised again in 1982/83 and in the following years, but they were still insufficient to cover the actual expenditures. In the 1970s, the rate of collection from the farmers in the Lower Seyhan Project area was very low but increased after 1982.[31] Since 1989 subsequent data has not been available, but the general trend of low collection rates has not changed. In 1993, for example, only about 10% of the assessed water charges had been collected.[32] Compared with the average Turkish collection rate of about 38.1%, the collection rate in the Lower Seyhan Project area is very low. Furthermore, it has to be considered that a part of the collected amounts are deferred payments of the previous years.[33] Tekinel et al. point out that the collection costs are not included in the figures, and that they actually exceed the collected charges. The DSI, however, bears part of the collection costs, which is not covered by its budget, by providing the vehicles for the local staff of the Ministry of Finance.[34]

1.3
The Water User Groups

With the completion of the stage I project at the end of the 1960s, the General Directorate for State Hydraulic Works initiated the establishment of Irrigation Cooperatives and Irrigation Associations, which were in operation until 1981. For a long time the farmers hesitated to join them because they believed that everything should be provided by the government, and the DSI operators worked like missionaries, as it was portrayed by the DSI's staff.[35] The Irrigation Cooperatives were weak in settling matters of conflict, the farmers did not pay the O&M costs, and irregularities (misuse) appeared with the administration of the collected water charges. The Irrigation Cooperatives thus remained few in number. After 1981 another type of water user organization was initiated, the Köy Tüzel Kisiligi (KTK)[36] which are based on the village administration units as the principal entities. From then until 1994 this type was widespread in the project area: from 115 water user organizations, 4 were Irrigation Cooperatives[37] and 111 KTKs. The

---

[30] DSI, 6.Bölge Müdürlügü 1986/87, p.11
[31] Other projects within the responsibility of the Sixth Regional Directorate show a collection rate of more than 50%.
[32] Tekinel et al. 1993, p.11
[33] See Tekinel et al. 1993
[34] Ibid.
[35] See DSI, Genel Müdürlügü, 17.Isletme ve Bakim Sube Müdürleri Toplantisi 1993, p.1
[36] Law No.293.4/101 (1981)
Köy Tüzel Kisiligi (KTK), literally translated, means 'village administration unit'. In the following text, the abbreviation WUG or KTK is used to address the type of water user group which is based around the village administration units.
[37] The Irrigation Cooperatives are associations of the beneficiaries, who elect their headman (a farmer, or an official person) and a council (five persons) through a two-thirds majority decision. The establishment of an Irrigation Cooperative, and its activities, are based on the Village Law No.442, and on the Civil Law No.743.

KTKs, or WUGs, supplied 95% of their irrigable area, which is quite a high number compared with an average all over Turkey of 46%.[38]

1.3.1
Mode of Establishment

To establish a Water User Group (i.e., a KTK), however, particular conditions must be fulfilled: a two-thirds majority of the beneficiaries in a village should declare their wish to establish a WUG. The beneficiaries should be residents of the village and should have their fields inside the village's boundaries. The DSI does not consider the establishment of a WUG in a village with many absentee-landlords. The farmers should be experienced in irrigated agriculture, or should seek experience from other villages which are already knowledgeable on the subject. A village assembly should have clarified the controversial question of whether they want to establish a WUG.[39] Villages which have only a small number of irrigable allotments are not encouraged to establish a WUG because the expenses for the WUG's contribution to O&M works would exceed their income. In the project area 21 villages have no Water User Group. In the Tarsus water subdistrict, for example, 15 villages have too little irrigable land, and their income would be too low when compared to their expenses.[40] An assembly of the beneficiaries, taking place in the village, clarifies the above mentioned conditions. If a majority of the beneficiaries who live in the village should declare their willingness to establish a WUG, an application will be made. Representatives of the state irrigation agency then join the assembly to reaffirm their wish.

1.3.2
The Water User Groups' Establishment

Each Water User Group is contracted with the DSI, and the contract is concluded with the mutual consent of the two parties. The 'Contract for the Takeover of Operation and Maintenance Works by the KTK' defines the particular responsibilites and duties for the Water User Groups in accordance with the DSI's establishment Law No.6200, article 2(k). It determines the subject, the Water User Group's responsibilities, the organization (personnel, equipment), the financing modality, the contract's duration, the modalities of notice, and the means of settling disputes. The contract, however, does not define obligations for the DSI. Separate protocols detail the maintenance and repair works to be done, and classify the work of the WUG which is expressed as a percent of the financial share of the total O&M costs, and how the rendering of accounts has to be executed. All contracts and

---

[38] DSI, 6.Bölge Müdürlügü, Isletme ve Bakim Sube Müdürlügü 1989, Annex 10
DSI, Genel Müdürlügü, 17.Isletme ve Bakim Sube Müdürleri Toplantisi 1993, Annex 1
[39] DSI, 6.Bölge Müdürlügü, Isletme ve Bakim Sube Müd. 1989; DSI, Genel Müdürlügü, 17.Isletme ve Bakim Sube Müdürleri Toplantisi 1993
[40] For example, if 16% of the total assessed water charges are given, then for 5 ha of cotton the village would get TL 304,000 (1991 prices) which is not enough to employ one person. (for details, see 1.4.4)

protocols are signed by the DSI's and the WUGs' representatives before they come into force. The contract lasts for one year, but the DSI can unilaterally cancel it.[41]

The headmen of the 111 WUGs (KTKs) in the Lower Seyhan project area are the village mayors, assisted by the village councils who take over the organization of O&M activities. 'Mayor' is the most important formal position of power in the village, and in this respect they are the people actually in contract with the state agency to execute services on behalf of the irrigation agency. They have to organize and monitor the O&M activities, and decide on personnel and equipment, and they have to administer the finances. The headman is entrusted with settling conflict among the irrigators in his village, and between the irrigators and the state agency, and he represents his WUG at court. As headmen of the Water User Groups, they are obliged to do their job properly. They have the overall responsibility to ensure the enforcement of the contracts and protocols; the impartial enforcement of the water allocation rules; to hire a sufficient number of qualified personnel and equip them well; to collect a share of the water charges to get the services financed, etc. As befits their office, they are committed to provide the irrigators, who directly pay the costs for the WUGs' operation and maintenance expenditures including the salaries of the headmen, with a satisfactory service.

1.3.3
Contributions to Operation and Maintenance

The WUGs hire ditch-riders, and the rule-of-thumb is one ditch-rider for 500 ha; their number should increase in times of water scarcity. The ditch-riders are persons who come from and live in the villages. They are well educated people with a lize degree, that is 10 years education, who, after they are employed by the Water User Group as ditch-riders, receive further training by the DSI to enable them to fully grasp the technicalities of their work. A ditch-rider supplies the service to the farmers of his village, and is employed for one year, or for one season. The ditch-riders contribute to operation acitvities, i.e., all activities to ensure the delivery of water from the secondary channels to the fields, monitoring the supply schedule of the irrigation agency, and collecting the pre-seasonal and actual water demand papers. They have to monitor the farmers' water withdrawal from the tertiary canals through siphons at the field outlets but they do not control it. They also have to look after the state of repair of the irrigation facilities, and report broken canalettes and the persons who destroyed them, and they should guarantee that rights-of-way are respected. A ditch-rider has to monitor his farmers' behaviour, and, if possible, the behaviour of the farmers of the neighbouring villages that share the same water course; he has to ensure an impartial water supply to all irrigators, i.e., following of the supply schedule. The ditch-rider is committed to his village and to his Water User Group, which employs him from one season to another. This is by far the strongest incentive for the ditch-rider to do his job properly. The irrigators can intervene if they are dissatisfied with the service he pro-

---

[41] For the contract and the protocols, see Annex 7

vides, as can the DSI, but its direct supervisory capacity on operation activities is low.

The effectiveness of monitoring the farmers' water withdrawal depends on the number of ditch-riders in relation to the area one ditch-rider has to serve (if it is vast or not); the number of farmers and outlets at field level; the time schedule (24-hour-basis or less), and whether they are equipped with adequate transport facilities. From the data available it can be concluded that some Water User Groups are well equipped with one or more cars or motorbikes, while others have only bicycles. Monitoring activities are by far the most important activity of a ditch-rider to ensure efficient water allocation, and to prevent 'overirrigation' caused by excess water withdrawals. In the Lower Seyhan Project area, the number of employed ditch-riders, the area they serve, and how they are equipped can only be roughly estimated. On a regional level, 342 Water User Groups hire 684 labourers for 155,364 ha, i.e., one ditch-rider serves, on average, 227 ha.[42] If one takes a similar figure for the relation between the ditch-riders and the area they serve in the project area, then one ditch-rider would serve 390 ha (115 WUGs with 230 ditch-riders, and 89,718 ha). This would not exceed the rule-of-thumb, but this figure does not coincide with the relation in reality. For example, in the village Sahinaga one ditch-rider operated 1,100 ha (and, probably, an additional number of temporarily employed labourers were hired). The WUG of the village Alifaki hired 2 labourers from May until September, one for 6 months, and another for the whole year, and irrigated 2,400 ha, plus 1,200 ha with second crops. This number doubled in 1989 when water was scarce. Despite the greater area one ditch-rider serves, both Water User Groups are perceived as good, or excellent, by the DSI personnel.

The contract is weak in defining what a sufficient number of adequate qualified personnel and sufficient equipment is, and the DSI cannot directly interfere with the WUG headman's decisions on how many ditch-riders, or labourers, are to be hired. But the annual field inspections by DSI personnel together with the WUG representatives exercise some control over the executed maintenance and repair works. An indicator for monitoring is the changing number of WUGs from year to year (some get brought to notice due to 'bad' work), and the degrading of some to a lower classification. Because of limited manpower, the WUGs contribution to maintenance and repair works is, however, not very great. They only carry out manual work, however, and do not participate in maintaining the drainage systems. They are obliged to execute the works before the irrigation season starts, which is not always the case.

There are innervillage commitments to rule-opposers, and the ditch-riders are to follow them. For example, poorer farmers with little land, and small labour forces, are allowed to withdraw water whenever they want. They are not forced to irrigate at night. This is also accepted by the DSI personnel. If larger farms withdraw water illegally, then the way they are treated depends on the authority of the headman and, if the events are reported to the DSI, on the DSI's intervention, and the imposed sanctions. The sanctions, if a rule opposer is detected, are not, however,

---

[42] Personal communication with staff from the regional O&M division of the DSI

harsh. The irrigator is warned, and if the illegal withdrawal is not stopped, his siphons will be destroyed. Water is not cut off. The ditch-riders, however, are not able to force the farmers and to execute sanctions. This is the duty of the WUG headman, and the DSI personnel only intervenes if a WUG wishes its mediation and if rule opposers are reported. The farmers themselves are actively involved, and they go to influencial persons for interventions, be it a DSI employee or the local police.

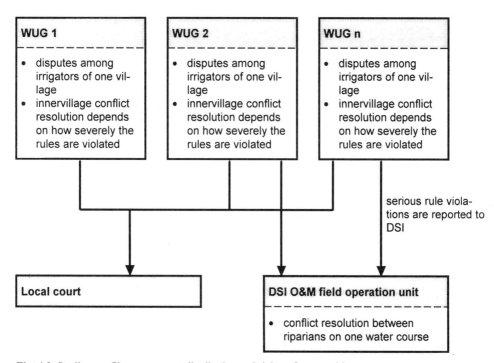

Fig. 4.2. Settling conflicts over water distribution and rights-of-way problems

The farmers keep in close contact with the DSI's staff pressing for better services. The buildings of the field operation units and the headquarters are easily accessible by road, and the housing of the employees is well known, and well frequented by the farmers. It is reported that the farmers do not hesitate to go to the house of the public employee, even during the night, if problems arise, and they do not hesitate to complain to whatever level of hierarchy is concerned, personally or through their representatives.

1.3.4
Financing the Water User Groups' Irrigation Services

If Water User Groups participate in O&M work, they receive a share of the totally assessed water charges as a monetary compensation for their work.[43] The extent of the share depends on a classification by the irrigation agency; this scale of classification ranges, usually, from 12% to 25% of the totally assessed water charges. If a WUG only participates in operation activities, the range is between 12 and 20%. If a WUG contributes towards operation, maintenance and repair works, then 16 to 25% are distributed. The law does not fix a category below 12%, but de facto 4, 8 or 10% are possible. A Water User Group usually receives a share of between 12 to 16% which is in the middle of the scale; for 'bad performance' 8%, and a maximum of 25% for 'good performance'. A WUG receives 25% if, for example, irrigation is done at night. This incentive was introduced to increase the overall water-use efficiency, and to compensate the WUGs for the higher expenses involved in night irrigation. (see Box 4.1.) If one uses the classification from 'good' to 'bad' as the range for qualifying the O&M activities, then the majority of the WUGs in the project area are classified as satisfactory by the DSI.[44] But this indicator is somewhat misleading: the DSI hesitates in reprimanding a WUG because it considers their positive effect on irrigation efficency, and without a WUG, the demand on the DSI's services would increase (but not necessarily the DSI's O&M budget). Therefore, the degrading of a WUG to a lower class seems, however, more rational to the DSI than to disband the WUG.

The headman collects the WUG's share of water charges directly from the beneficiaries. From their share of the water charges, the WUGs have to finance the salary of the headman, the salary of the ditch-rider(s) and his (their) national insurance, the wages for the seasonal workers, the equipment, the vehicles, which can be bought or leased, and fuel, etc. In accordance with the contract, the WUGs are obliged to hire a sufficient number of adequately qualified personnel, and to equip the personnel with the necessary transportation, such as cars or motorbikes, etc. What sort of equipment or vehicles the WUG is to buy, or to lease, and how many are required, if at all, is the headman's decision. The WUG headman can, within certain limits, freely decide on the amount of money to spend on O&M works. The DSI does not control or directly interfere in such matters. But the collected share,

---

[43] Unlike in many other large-scale irrigation schemes throughout the world, where the irrigators contribute to operation and maintenance acitvities in kind, the irrigators in Turkish schemes do not contribute with labour.
See Annex 6: Annual Decree on Water Tariffs for O&M and investment, and Annex 7: Contractual agreements between DSI and the Water User Groups

[44] The WUGs in the Yüregir Plain are classified as follows: In 1987, 3 WUGs received 8%, 1 WUG received 12%, and 58 received 16%. In 1988, 2 WUGs received 12% and 50 WUGs 16%. Most WUGs in the Tarsus Plain, i.e. 57, get 16%, and, in 1987, four in the Tarsus Plain were reduced to 12% for 'bad' work. (DSI, 6.Bölge Müdürlügü, Dogankent Isletme ve Bakim Basmühendislik 1989; DSI, 6.Bölge Müdürlügü, Ismailiye Isletme ve Bakim Basmühendislik 1987,1988; DSI, 6.Bölge Müdürlügü, Yenice Isletme ve Bakim Basmühendislik 1987,1988)

and all expenses, are strictly controlled by the DSI, the administrative superiors, the public prosecutor's office, and the irrigators to prevent misuse.

Box 4.1. Water User Groups in the project area[45]

The WUG of the village Sahinaga (Yüregir Plain)
The Water User Group of the village Sahinaga was established in 1968. It is perceived as a 'good' example by DSI personnel. The village has 60 households with 302 inhabitants, and 100% of 1,100 ha of irrigable land is irrigated. Sahinaga receives 16% of the assessed O&M costs. Since 1968, the WUG has financed the construction of a road, the electrification of the village, and a school building. A foreman (a relic of the Farmers' Extension Service) controls the water application, and it is said that far too much water is applied by the farmers. The Water User Group has some trouble with the neighbouring village which uses too many siphons (3 more than allowed), and thus too much water.

The WUG of the village Alifaki (Tarsus Plain)
The Water User Group of the village Alifaki was also established in 1968, and is perceived as a 'very good' one (25%). The village is, with 1,500 residents, the second largest in the project area. The WUG serves 2,400 ha, plus 1,100 ha which are second cropped. It is equipped with 3 motorbikes and one car. 50% of the farms, which are larger ones, irrigate at night, while the smaller farms irrigate only in the daytime. 20% of the farms do not pay the O&M costs due to their small size. Since 1968 the WUG has, with the surplus, financed roads, a school building, a burial ground, the connection to telephone and electricity networks, and a mosque.

A particular regulation provides a strong incentive for the headman to guarantee that a WUG's contribution towards O&M remains, at least, 'adequate' as perceived by the DSI's staff: the village administration unit is legally allowed to spent a part of the collected irrigation charges for village projects. There is, usually, a surplus because the WUG's personnel is cheaper than the DSI's. A person employed permanently for operation services costs TL 2,5 million/year (WUG-employed) or TL 9-10 million/year (DSI-employed). If maintenance and repair works are also considered, the advantages would be much greater.[46] As an incentive to the headman, the remaining surplus depends on the DSI's recognition that the WUG's work, including that of the ditch-rider, is 'good'. It is in the interest of the headman to keep the executed works, in particular that of maintenance, at a level where a surplus remains. This surplus can be used for village projects, such as the construction of roads, bridges, mosques, school buildings, electrification, telecommunication networks (telephone), or the construction of a drinking water well. According to Turkish Village Law, villages are self-governing autonomous local administrations. Their revenue is composed mainly from local sources, and 50% of the Iller Bank's net profit has to be spent on buying property for villages. Law No.4759, Art.19, defines these properties as being of the kind that allow the villagers to work on/with them. The main local source of revenue was 'Imece', i.e.,

---

[45] Personal communication with WUG headmen 1989
[46] Personnel communication with the director of the O&M division of the General Directorate for State Hydraulic Works

a compulsory contribution of labour to public works and services by the villagers, interpreted as mutual help among individuals. If a family could not contribute, it had to pay others to do its work. The cash revenue consists of 'Salma', a special tax imposed on each family in the village which is decided on by the Council of Elders. The annual amount is specified by law, depending on the economic situation of the family. Other cash sources are fines, rental revenues (mainly the rents for village-owned fields) and charges to be paid for certifying documents. Due to the fact that the villages have to pay in cash for most of the works to be done, 'Imece' is of little importance. 'Salma', the basic element of revenue, often cannot be, or is not collected.[47] Thus, a village has little or no revenue at its disposal, and, therefore, the share of the collected water charges plays an eminently important role because the surplus is an important source of finance for village projects. It is a great incentive for mayors to take over O&M responsibilty, and to make the irrigators pay their charges. On the other hand, it is an incentive to save money for village projects, and to keep the O&M expenses at a limited level. If a village gets roads, or electricity, a telecommunication network, a school building, or a mosque, etc., then the village mayor, as a politician, collects 'good points' for the next election. In 1987 and 1988, all Water User Groups in the project area saved a considerable amount of money for village projects. In the Yüregir Plain, approximately 50% could be withheld. The Water User Group of the village Sahinaga, for example, collected about TL 4 million in 1988; the headman's salary was TL 0.8 million, and the expenses for maintenance and repair works TL 1.5 million. The remaining TL 1.7 million constituted the village budget. In 1987, in the Tarsus Plain, the withheld surplus was about 40%, while it remained at only 10% in 1988.[48]

The collection of water charges, however, faces some difficulties although the collection rate ranges from 50 to 100%, which is higher than the rate that the irrigators pay to the Ministry of Finance (i.e., about 30 to 10%). Water charges are collected after the harvest, and the first steps that the farmers make after receiving their money is to pay back any debts they may have to a bank, or make investments. The collection rate depends on farm sizes and farm structures (share-cropping, leasing), and it is reported that the smaller farms and the share-croppers are more unwilling to pay because they have high production costs and thus have a low profit. Tough kinship relations, and striking political struggles have negative impacts on the collection rate.[49] The collection rate could be much higher if non-payment was sanctioned, and if the collection procedure was changed. For example, if the irrigators could pay their share of O&M costs by instalments to the Water User Group, the first quota would be paid for operation works at the beginning of the irrigation season, while the second quota, for maintenance and repair works, would be paid after harvesting, when maintenance and repair work starts. If the

---

[47] See Tug 1975

[48] DSI, 6. Bölge Müdürlügü, Ismailiye Isletme Mühendisligi 1987, 1988; DSI, 6. Bölge Müdürlügü, Dogankent Isletme ve Bakim Basmühendisligi 1989; DSI, 6. Bölge Müdürlügü, Yenice Isletme ve Bakim Basmühendisligi 1987, 1988; Interviews 1990

[49] See DSI, Genel Müdürlügü, 17.Isletme ve Bakim Sube Müdürleri Toplantisi 1993, p.12

first quota for operation is not paid, it is more likely that the threat of a water cut off at the beginning of the irrigation season would prove to be one more way of persuading the farmer to pay.[50]

1.4
Performing Operation and Maintenance, and the Contributions to High Groundwater Levels and Salinity Caused by O&M Activities

The point of issue in this section is to reveal whether the operation activities guarantee that water supply services are adequate and reliable, whether they instigate, or prevent excess water inputs; and to discuss, finally, the contributions to high groundwater levels and salinity made by the maintenance of the scheme.

1.4.1
Performing Operation, and Contributions to High Groundwater Levels and Salinization

1.4.1.1
The Procedures for the Planning and Daily Implementation of the Water Supply

In state-owned and -managed large-scale irrigation projects in Turkey, all farmers within a service area have, in principle, non-transferable irrigation water user rights for which they have to apply every year before the irrigation season starts. The state agency decides on the rules for water allocation, e.g., to whom and at what time a certain amount of water is going to be delivered, and implements the necessary operation activities to supply the farms. The water distribution method in use is a 'semi-demand' system, which means that the irrigation water is made available to the farmers within a few days of their request.[51] Fixed amounts of water are released per hectare, the calculations for which are based on the crops' water requirements and evapotranspiration coefficients, precipitation, and pre-assumed water conveyance efficiencies and on-farm efficiencies. The calculation method in use is the Blaney-Criddle-Method. The farmers have a free choice over the cropping pattern, influenced by state-supported output prices, with their response to the market being the determining factor, and the DSI does not interfere in cropping decisions.[52] In years of water scarcity, when the stored water is insufficient to satisfy the requests of all farmers, those farmers who announced their demand first, are served first, leaving other farmers without water, and without compensation.

---

[50] This procedure has not yet been introduced but has been discussed in a DSI workshop on WUG issues. (see DSI, Genel Müdürlügü, 17.Isletme ve Bakim Sube Müdürleri Toplantisi 1993, p.13)

[51] See Box 2.1. Water distribution methods

[52] Politically, it is, nevertheless, intended that the water charges, which are assessed per hectare and crop, would exercise some influence on cropping decisions.

For calculating the seasonal water supply, the agency needs to gather information about the intended cropping pattern throughout the service area. The DSI introduced a procedure for gathering data to accomplish the seasonal supply plans: approximately one month before the irrigation season starts, a 'General Irrigation Plan' is required from each farmer who has to state his intended crop pattern and the area to be served with irrigation water. The farmers are obliged to pass on their request to get the seasonal irrigation water user right. Two days before the farmers want to irrigate their fields, they announce their actual water demand with a 'Water Demand Paper' declaring which crops have been planted, the size and location of land, the flow rates required, and the desired day and time of water release. The Water Demand Paper is passed on directly, or through the Water User Groups, to the DSI's district technicians. If they use irrigation water without applying for it, the Turkish law provides a single 10% fine on O&M costs. If the farmers do not follow this formal way, or provide false information, user rights can be denied, and the DSI can refuse water deliveries in accordance with the Water Charges Assessment Decree. (see Annex 6, Art.12) The daily supply schedule is published in the villages' coffee-houses to keep the farmers informed about the amount of water, the number and size of siphons to be used, and the time and duration of release. The farmers are obliged to follow the schedule, and the publishing of the daily schedule is also an effective means for mutual control amongst the irrigators in one village.

The DSI's technicians, together with the ditch-riders of the Water User Groups, operate the regulatory networks, and they have the duty of monitoring and strictly enforcing the water supply schedule at the farm outlets (e.g., the time, and the number and size of the siphons), and in seasons with low water conditions it is expected that the supply schedule is more strictly monitored by those responsible. The DSI regards the enforcement of the water supply schedules as an important factor for improving the overall water-use efficiency, and for restricting excess water withdrawals by the farmers. Illegal water tapping,[53] e.g., withdrawing more water and opposing the time schedule, should be reported to the DSI officials who are allowed to impose fines and execute sanctions. Siphons can be destroyed, but water is not cut off.

1.4.1.2
Evaluating Operation

The evaluation of the operation activities draws heavily from interviews with the DSI's staff, various DSI publications, and from ex-post evaluation reports of the FAO.[54] The reviewed literature only sometimes provides empirical proof of whether the operation activities actually cause water distribution problems and

---

[53] The term 'illegal' tapping is used whenever the supply schedule is violated, e.g. the assigned amounts of water are exceeded, and/or when the timing is opposed.

[54] FAO 1987, 1988; DSI, 6.Bölge Müdürlügü 1986/87, 1988a, 1989a; DSI, 6.Bölge Müdürlügü, ASO Subesi (1989-90, 1990-91, 1991-92, 1992-93, 1993-94); DSI, 6.Bölge Müdürlügü, ASO Subesi 1990; DSI, Genel Müdürlügü 1995; DSI, Genel Müdürlügü, Isletme ve Bakim Dairesi Baskanligi 1989

contribute to high groundwater levels, or not. The results can, therefore, only be assessed in the following classifications: (*) Excess water inputs into the area are empirically observed, and can thus be proven; (*) the operation activities generate distribution problems, and/or it is likely that they induce, or cannot prevent excess water inputs. Various issues of particular concern are specified, some of which refer to the principle definition of the farmers' user rights and whether they are defined in such a way to limit uncertainty in supply; the planning and daily implementation of water supply is evaluated whether the water distribution is adequate, reliable and predictable, and what causes high groundwater levels and salinity.

Farmers' user rights to irrigation water

It is a common feature that a large number of farmers do not apply for water, but, none the less, practice irrigation, although it is a prerequisite to obtain the seasonal user rights. In 1989, for example, only 76% of the farmers who wanted to irrigate passed their plans on. According to the Water Charges Assessment Decree, user rights can be denied and water deliveries can be refused, but this sanction is not enforced. Fines are imposed which only amount to 10% of the O&M costs, irrespective of the date of payment, and, therefore, they do not encourage the farmers to participate in planning. The abjudication of seasonal user rights is not politically desired, and the fines, if they are imposed, are rarely collected. As was previously mentioned, if fines are actually collected or not depends on the Ministry of Finance's staff, and there is little motivation for effective collections as the administrative costs of collecting fines are higher than the money received. The DSI has no responsibility or control over this issue, it is not even informed by the staff of the Ministry of Finance. However, without the farmers' applications, the planning process to guarantee adequate water supply becomes increasingly difficult and inaccurate, if not impossible and, therefore, the DSI considers the farmers' user rights even when they have not been applied for by using their data from the previous year.

In years of normal water availability, water distribution follows the farmers' requests, and water is delivered when they need it most. In years when the available water does not suffice for the whole service area, water is provided according to the principle „Who applied first, is served first". This means that some farmers are deprived of their user rights even if they have applied for them, and they are excluded from the use of water after they have decided on their cropping pattern. These farmers might be deprived of a whole year's income without any compensation. Because the farmers have no access to data, and no control over the designation of user rights, insecurity increases on the side of the farmers. In the years 1989 and 1990, for example, the irrigation agency officially announced a 'water shortage'. The farmers did not believe that this was true. This situation, e.g., the designation of user rights in times of water scarcity, and the lack of access to information data, only created incentives for the farmers to ignore the supply schedule, resulting in distribution problems.

The capacity of the delivery system was designed to provide water for a preestimated cropping pattern which had never been achieved. It was estimated that cotton, a high water-consuming crop, would be planted on 35% of the project area

In 1966, cotton was planted on 97.4% of the project area, and every year during the 1980s it almost always reached 80%.[55] Although the water available was sufficient from a general point of view, in times of peak demand the requirements of all riparian farmers exceed a channel's capacity, especially if all farmers had planted high water-consuming crops. This then means that the farmers at the lower end of the system receive their water several days later than the date they had originally requested. The technical system in combination with the semi-demand delivery method thus cannot always guarantee water deliveries being on time, and this problem occurs irrespective of rule enforcement. Some farmers face the threat of not receiving their water on time, or even at all. For both situations, i.e., low water conditions and times of peak demand, a proportional reduction of the water shares, spreading the risk and making supply more predictable, would reduce distribution problems, with higher operation and enforcement costs.

Some farmers have no direct access to the delivery system because of the wide spacing of the tertiaries. The problem increases due to progressive land fragmentation caused by the inheritance law. But, nevertheless, these farmers have user rights, and their demands are considered and reflected by the actual supply scheduling. Legally, rights-of-way have to be respected by all farmers, but this, unfortunately is not always the case. If their rights are not enforced, they are deprived of income, and compensations would be necessary. Strict rule enforcement, however, and the imposition of sanctions, or the extension of the technical system, could guarantee access. Just as rule enforcement creates high monitoring and enforcement costs, particularly for the Water User Groups, the extension of the technical system which is already relatively dense, would increase investment and maintenance costs.

Despite the fact that the farmers have no user rights to irrigation water in the area of the stage IV project, they illegally practice irrigated agriculture, withdrawing water from the main delivery channels, if possible, and, in particular, from the drainage channels with their highly saline water. The irrigation agency should have prevented the farmers from irrigating, but could not due to high monitoring and enforcement costs (in this part of the project area, there are no Water User Groups). The farmers' practices, however, have become legalized, and water charges have been levied.[56] The consequences of this action, though, are saline water inputs into the lowlying regions of the project area where groundwater levels and saline soils are already critical, and where proper drainage systems are lacking.

Planning and daily implementation of water supply
The universal method for calculating the crops' water requirements does not allow for specifications within the service area, e.g., the soil types, their drainage conditions, and the irrigation method. The calculated and released amount of water per

---

[55] See Annex 3: Crop patterns
[56] In this area, the farmers had to pay 35% of the normally assessed charges until 1993; since 1993, they have to pay 20%. Whether the charges are collected is doubtful; the agency has no control over water withdrawals and no information whether the lands are irrigated or not.

crop and hectare has been discussed and questioned for a long time by the agricultural research centre of the General Directorate for Rural Services, and by scientists of the Çukurova University. Based on field experience programmes for cotton, soybean, maize, citrus and melons, conducted in all local soil series, it has been concluded that, with cotton for example, the amount of water applied could be reduced by about 25% without significant impacts on yields (under constant irrigation techniques, i.e., furrows). A similar reduction is assumed for other crops.[57] There is no mechanism to utilize this potential for water savings as a means for groundwater table control as the General Directorate for Rural Services do not have the power to enforce such reductions, even though the technical assistance for water-use at on-farm level and the protection of soils legally rests with the agricultural agency. The DSI, while commanding the delivery of water, has taken over control, although it is, in the view of the GDRS, not legally entrusted to do so. This conflicting issue originates from the simple fact that commanding water deliveries through the DSI's activities are more influential than the instruments available to the agricultural agency. The General Directorate for Rural Services can only give advice to farmers through demonstrations. Political decisions were missing, and cooperation/ coordination between the state agencies was not forthcoming. The consequences of unadjusted water releases are, however, excess water inputs into the area caused by the main system's management, which increases drainage requirements at on-farm level.

A great deal of information and coordination is required for the daily planning of the water supply due to the farmers free choice of cropping patterns. Difficulties occur when not all current Water Demand Papers are passed on. The irrigation agency tries to overcome these difficulties by using the farmers' data from the previous year. It is reported that the larger, well managed farm units usually announce their demands on time. However, farmers who did not apply for water still practice irrigation, overtaxing the supply and the scheduling, which results in delays in supply for downstream farmers.

There is always some discrepancy between the intended crop and that which is in effect planted. Some of the farmers change their decision on crops (e.g., other crops with higher water requirements are planted than those announced), while others do not irrigate at all. In 1990, for example, it was announced that an intended 157,237 ha (incl. second cropping) would be irrigated in the project area, while in actual fact only 99,475 ha were irrigated. Usually, the released water, based on the announced demands, exceeds the amounts actually withdrawn. The operational water-use efficiencies were 45% (1984), 49% (1985), 37% (1986), 44% (1987), 47% (1988), and 67% (1990), because the farmers applied for water, but did not use it. In 1990, for example, 300 million cubic metres of water were released in excess of requirements. As a result, the high content of sediment within the 300 million cubic metres of water burdened the irrigation networks with about 4 million cubic metres of silt deposition, and thus created higher maintenance costs. Supply problems may also occur at a later point in time, and for water districts downstream.[58]

---

[57] See Tekinel and Kanber 1988
[58] See DSI, 6.Bölge Müdürlügü 1989a, 1989b

The crops' water requirements, evapotranspiration, water availability, precipitation, water conveyance efficiencies and on-farm efficiencies are taken into account when calculating the gross amount of water to be released. While the water availability can be estimated more precisely due to the existence of measurement stations, the data on water conveyance efficiencies and on-farm efficiencies are incomplete. The irrigation agency usually reacts to inaccurate information by feeding more water into the delivery system in order to satisfy the farmers' demands. If water losses occur due to the poor maintenance, or unrepaired leaky channels, the irrigation agency again counteracts these losses by feeding more water into the delivery systems. Precise data on locally varying water conveyance efficiencies and on-farm efficiencies could reduce water deliveries but would produce extraordinarily high information costs, with information still being inaccurate. More effective means of improving the operational efficiency are better maintenance, and incentives for the improvement of on-farm efficiencies. The consequences are, however, that more water is released than needed, and unabated seepage from the channels.

In some years, due to dry weather conditions, the irrigation season had to start earlier. In February and March 1989, for example, cereals had to be irrigated, and at that time neither the general nor the actual demand papers had been collected. In other years, dry weather conditions in May and June forced the farmers to irrigate cereals, and they applied for water (usually, the precipitation is sufficient for cereals). Changing their minds with the weather, they decided not to use it, and it discharged, unused, into the drains. Natural events pose high coordination costs to adjust water supply, and whether sufficient water is supplied, or not, depends on the cooperation/coordination between the farmers, the Water User Groups, and the management unit, to overcome any information deficits, and thus uncertainties for the farmers.

The flow rates in a large number of secondary channels in the most intensive irrigation months of July and August, were below the channels' capacity, creating an artifical water shortage. It is presumed that the importance of making accurate measurements and keeping reliable records is not fully understood by the staff in charge of measuring and recording flow rates. More training for the staff was thus found to be necessary.

The technical system is calculated to deliver sufficient water on a 24-hours-basis but is used only 15 to 17 hours per day, because the farmers do not irrigate at night. The 24-hour supply-scheduling is one of the main reason why farm operators, or their labourers, oppose the schedule, and illegally tap water. Proper irrigation practices are difficult to execute at night, and labour forces are more costly. Monitoring at night, which is the duty of the ditch-riders, is also extremely difficult. Positive incentives for night-time irrigation, such as the provision of higher shares from water charges for the WUGs, show little effects when water is abundant. Water discharges unused into the drains, and supply problems may occur at a later point in time. Only under low water conditions do larger holdings use the night flows, while the smaller farmers are allowed to irrigate in daylight. Another important aspect which derives from the technical layout of the delivery system is that if only a few farmers need water, a particular flow has to be upheld in the main branches in order to feed the tertiaries, and most of the water discharges unused into the drains. The technical layout of the delivery networks in combination with the semi-demand distribution method may cause a water deficit at a later point in time, or in downstream districts.

1.4.2
Performing Maintenance, and Contributions to High Groundwater Levels and Salinization

1.4.2.1 Maintenance Programmes and Activities

In the DSI's annual reports on groundwater and salinity, the causes of rising groundwater levels and recommendations for their alleviation are documented, and they are, as far as they concern the DSI's activities, transformed into annual maintenance programmes. These reports also comprises recommendations for improvements of the on-farm development works, over which the DSI exercises no control. The responsibility for the maintenance of the drainage networks at on-project level, i.e., the main, secondary and tertiary drains, rests solely with the DSI. According to the DSI's manual, the main drainage channels should be cleaned every four or five years, the secondary drains every three or four years, and the tertiary drains every two or three years. Maintenance and repair works of the project's delivery systems are the joint responsibility of the DSI's Lower Seyhan Branch and the Water User Groups. The Water User Groups participate in manual maintenance works in the secondary and tertiary delivery systems while repair works are carried out on a voluntary basis. The DSI's staff is responsible for mechanical works, and for chemical and biological eradication programme activities. The delivery channels maintenance work comprises silt and weed removal; maintaining the regulatory devices to keep them in working order; the repair of leaky canalettes, and the replacement of broken ones.[59] Local maintenance and repair needs of the deliveries are identified in annual field inspections made by the DSI's staff and the Water User Groups. They both, then, determine the manual works which should be carried out by the Water User Groups before the irrigation season starts. Available reports indicate that the WUGs are involved, foremost, in sediment removal.

1.4.2.2 Evaluating Maintenance

According to the reviewed reports, the maintenance of the delivery systems seems to pose less problems when compared with the maintenance of the drainage systems. The DSI has set first priority for keeping the water distribution networks operational. Desilting and weed removal, repair and replacements are executed with the DSI's own forces (i.e., personnel, equipment, etc.,), and all necessary

---

[59] The conveyors, the main, and secondary channels are concrete lined, and some of the secondary, and all tertiary channels, are canalettes. Canalettes consist of single prefabricated and elevated cement shells.

expenditures are maintained as much as possible.[60] The high water conveyance efficiencies can serve as an indicator for the performance: within the main canals it reaches 92 to 95%, the secondaries average at 90 to 95%, the canalettes at 88%, and the tertiary canals between 90 and 94%.[61] A sample survey, conducted in 1985 in the stage II project area, however, pointed to the farmers' problems with the maintenance activities in the delivery systems. The respondents claimed that leakage and flooding damaged the crops; repairs were not executed under the supervision of skilled personnel; the capacity of some deliveries was regarded as insufficient because canals were not kept clean. Losses of water had been reported by a significant number of respondents.[62] The available data for the last years show that the executed maintenance programmes of the delivery networks, although reduced, has remained on a higher level than the maintenance of the drainage networks, which declined from 71% in 1990 to 32% in 1994.[63] (see Table 4.9.) Until the middle of the 1980s budgetary allocations for maintenance were not enough, but they have again been curtailed since the mid-'80s, and much more sharply since 1990. Table 4.9. indicates, however, that between 1990 and 1994, the planned programmes for canal cleaning (i.e., desilting and weed removal,) have been nearly maintained.

Table 4.9. Executed maintenance as a percentage of the planned programmes[64]

Issue	1990	1991	1992	1993	1994
Delivery canal cleaning	86	81	87	86	76
Repair of concrete-lined delivery canals and canalettes	54	45	39	39	32
Canalette replacement	86	51	29	39	n.a.[a]
Drainage canal cleaning	71	61	51	82	32

[a]   no data available

[60] DSI, Genel Müdürlügü, Isletme ve Bakim Dairesi Baskanligi 1995; DSI, Genel Müdürlügü 1995
[61] Yavuz (1985) and Önes (1976), quoted from FAO 1988, p.25; DSI, 6.Bölge Müdürlügü 1986/87
[62] The World Bank 1985, p.35
[63] See DSI, Genel Müdürlügü, Isletme ve Bakim Dairesi Baskanligi 1995, p.8; DSI, Genel Müdürlügü 1995, p.41
[64] Source: DSI, Genel Müdürlügü, Isletme ve Bakim Dairesi Baskanligi 1995, p.8; DSI, Genel Müdürlügü 1995, p.41

The budget cut-backs concerned, foremost, the maintenance and repair activities for which cash allocations were needed, for example, the repair of structures and canal lining, although, within the maintenance envelope, the upkeep of the deliveries gained priority. As deferments have become indefinite, the networks are now in a state of disrepair. Since 1990, waterlogging along the watercourses are well-known phenomena. Farmers go to court because of crop losses caused by waterlogging from unrepaired channels. In most of the cases, the DSI loses the case, and compensation has to be paid.

The intervals of the drainage maintenance programmes, which are set by the DSI, are longer than the time intervals required for a properly maintained system, and even these intervals are not ascertained. The DSI can only clear the main drains once every six years, rather than every three or four years as required.[65] The annual maintenance programmes are only partly executed, and delays frequently occur.[66] Due to the inadequacy and shortage of machinery, and due to the low labour productivity, the DSI has accumulated a large backlog of drainage maintenance work. Since the early 1980s, however, the DSI has applied chemicals into the drainage systems as a response to the striking weed infestation, which results from ineffective mechanical cleaning. The unstabilized banks of the main drainage channels are treated with selective herbizides one to three times a year. The drainage channels - the bank area and the channel water - are treated with total herbicides.[67] It is, however, reported that chemical eradication programmes have been insufficient to solve the maintenance problem in the drains. Biological methods to reduce the growth of hydrophytes in the main channels, like using a kind of carp (Ctenopharyngodon idella), are rarely applied as the fish are caught by the seasonal workers. The poor state of the surface drains affects the closed-conduit drains at on-farm level, which are sometimes submerged. The outlets of the on-farm drains are frequently blocked or broken, either by the farmers or due to poor maintenance, and the drainage systems cannot effectively lower the groundwater levels. Siltation and weed infestation reduces the flow rates, and creates favourable conditions for mosquitos and other insects.

As the poor state of the main drainage networks affect the on-farm drainage systems, they themselves are in poor state. Under the present conditions, where investments in on-farm works are financed by the state, the farmers have no incentive to participate in the maintenance of the on-farm installations, and the maintenance of the subsurface and surface drains has been omitted. The state agency is not able to maintain these extensive infrastructures, and it is not legally entrusted to do so. The farmers, as the non-owners of these installations, do not feel respon-

---

[65] FAO 1988, 1987; DSI, 6.Bölge Müdürlügü 1986/87; Interviews 1989; TU Berlin 1992
[66] FAO 1988, p.30
[67] To clear weeds in the bank areas 2,4 D-Amin (0.8-1.6 kg/ha) is applied, and Glyphosat (4.8-7.2 kg/ha) and Simazin (5.0-7.5 kg/ha) to clear woody growth. For the treatment of the hydrophytes Glysophat and rarely Dalaphon (CKW) is used on the water surface. The application of Dalaphon is decreasing because of the high dosage (25-42 kg/ha) and the short term effectiveness. For hydrophytes and algea Endothall Alkylamin (7.0 ppmw), copper sulphate (3.0 ppmw) and copper ethanolamine (1.0 ppmw) is let into the channel water. (see TU Berlin 1992, pp.111-112)

sible for the maintenance of the 'state property'. It is probably not only that the farmers do not feel responsible, but that they do not want to risk anything, whatsoever, on state property.

The main drains are not only used for agricultural runoffs, but also for industrial and communal effluents. Almost all industries (e.g., textile, paint, plastic bottle, food, grease, oil, etc.,) discharge untreated waste water into the main drains. Slaughterhouses, chicken factory-farms, and vegetable wholesale markets use the main drains as dumps. The untreated waste water of half a million residents of the city of Adana is also released into a main drain. The discharged volume does not actually create a capacity problem, but with the population, and water consumption per capita constantly increasing, capacity problems will emerge in the near future.[68] These usages increase the maintenance needs, but neither the industries nor the city of Adana contribute to maintenance costs, and they are not charged for the use of the drainage systems.

1.5
Conclusions on the First Public-Farmer Setting

The DSI was able to manage many distribution problems, because the water supply in the project area is favoured by natural, technical and institutional conditions. Water is usually abundant, and only in 1988 and 1989, for the first time since the operation started, were water conditions low. Due to the fact that the fourth stage of the project area, or about 40,000 ha, is not yet connected and supplied with water, all available water can be used in the three stages already operating. A legally bound decision has been made that irrigation needs have priority whenever a competing situation appears between irrigation and energy generation, i.e., no water deficit has occurred due to energy generation.[69] The semi-demand distribution method provides water as demanded by the farmers. This method can increase on-farm efficiencies, compared with other distribution methods, because water is delivered at the farmers' request and when they need it most. The management of this system is, however, increasingly complicated but in nearly all villages Water User Groups participate in operation activities. The concept of the Water User Groups based on the village administration units has provided, however, benefits for all parties involved:

* The DSI hoped to increase the irrigation ratio,[70] and the General Directorate considers that with the Water User Groups this ratio ranges between 70 and 83%, and that, generally speaking, the Water User Groups have had a positive

---

[68] Personal communication with the director of the planning department of the Regional Directorate of the DSI 1995

[69] In 1956, when the hydroelectric power plant was rented out for 49 years to the Çukurova Elektrik Sharehold (CEK), a contract was signed between the DSI and CEK agreeing that irrigation has first priority. The DSI is assigned to control the CEK, and the contract can be unilaterally noticed by the DSI, if CEK does not deliver the guaranteed water.

[70] That is the relation between the irrigated and irrigable area.

impact.[71] Without the support of the Water User Groups, the service would have to be provided by the state agency. If no other organization would contribute to O&M works, the DSI would have been obliged to enlarge its staff, and, of course, its budget, because the O&M works exceeded, by far, its personnel capacity. However, without Water User Groups the DSI could not manage to deliver water as requested by the farmers.

* The Water User Groups' participation in O&M works lowered the burden on the general treasury, firstly because it reduces the O&M budgets, and, secondly, because the government saves money on village development activities.

* The financing principle provides the most striking incentive for the village mayors to take over O&M responsibilities; they receive an additional income and they can save money for village development projects due to lower labour costs.[72]

* The farmers as village residents benefit from the undertaken projects which are, in fact, to the benefit of all villagers, and financed only by those farmers practising irrigated agriculture.[73] This commitment, however, seems to cause no problems. As farmers, however, they do not gain a monetary benefit if a WUG participates in O&M works, and they do not save money from the classification principle because they have to pay the total assessed water charges.

The Water User Groups have shown some disadvantages within the publicly run projects. The main disadvantage, as perceived by the DSI's staff, is that they had „only partial responsibilty such as cleaning and minor repairs of tertiary canals, and their level of commitment for the management is not as serious".[74] The WUGs faced difficulties associated with the fact that their boundaries are based on settlement rather than irrigation boundaries which creates difficulties in operating and maintaining canals that cut across village boundaries. The main disadvantage, however, was that there is a lack of direct farmer involvement, and that the irrigators had no institutionalized control over the WUGs' operation and maintenance activities.[75]

As a result of their institutional setting, the Water User Groups have not been successful in enforcing the water allocation rules, and they could not prevent excess water withdrawals by the farmers. To guarantee, or, at least to increase the probability of rule enforcement, more ditch-riders would be necessary. The low number of ditch-riders employed by the WUGs is, however, the result of the incentive system because increasing their numbers would reduce a village's benefit. The same holds true if one considers that more personnel for maintenance and repair activities should be employed because if the number of employed personnel would increase, the incentive for the village administration to contribute towards maintenance works would, at the same time, decrease.

---

[71] DSI, Genel Müdürlügü, Isletme ve Bakim Sube Müdürleri Toplantisi 1993, p.11
[72] Personal communication with the director of the O&M division of the General Directorate for State Hydraulic Works
[73] This can be regarded as indirect taxes; income taxes have been levied only since 1984.
[74] Mohamadi and Uskay 1994, p.10
[75] See DSI, IIMI and EDI 1996

The headmen of the WUGs, who are at the same time the village mayors, have, however, a difficult role to play. As politicians elected by the majority of the voters, some of their targets conflict with the office as headmen of the Water User Groups. As elected village mayors, they have responsibilities towards their voters. As village mayors their main legally defined duty is to secure peace and order within their villages. They represent their village as a unit to the outside world, and they act as agents of the national government. The mayors are usually members of political parties, or, if not, seemingly influencal persons who enjoy the majority's esteem. Sometimes, they are persons with less political influence, and elected by strong kinship groups, who might then interfere in village affairs.[76] As politicians they are interested in getting reelected, and to preserve their reputation as 'good village representatives'. According to the contract, the headmen have to contribute towards settling conflicts among the irrigators, and between the irrigators and the state agency, and towards the collection of water charges. If they force irrigators who are unwilling to pay their water charges, then it is likely that they will lose voters. If they enforce the rules, which includes the detection of rule breakers and, in some cases, their notification to the 'state', then the same can happen, and they will probably lose the support of the kinship-leaders in villages with strong kindred relationships. If there were conflicting objectives and interests between their duties as elected village mayors and appointed WUG headmen, they almost considered their political interests. The amalgamation of the two functions could explain why the village mayors are regarded as 'weak' persons with regard to conflict settlement, rule enforcement, and, in some cases, the collection of water charges. The DSI characterizes them as peaceable persons.[77] This means that the enforcement of the supply scheduling and the sanctions on rule opposers, and the collection of water charges, in particular, have remained difficult tasks not only because of too few personnel.

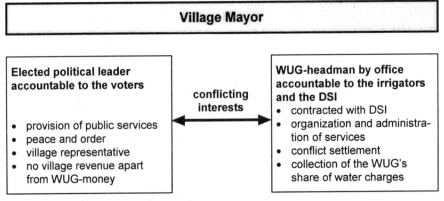

Fig. 4.3. The conflicting role of the village mayors as elected political leaders and headmen of the Water User Groups

---

[76] See Leder 1979, p.96; Dabag 1988
[77] See DSI, 6.Bölge Müdürlügü 1986/87, p.10

1. High water inputs have accrued from the main system's management.

On the management level, the major factor conducive to high groundwater levels and waterlogging are the calculated and released amounts of water per hectare and per type of crop, which could be reduced by approximately 25% without significant impacts on yields. The higher than necessary water inputs which have been an essential part of the DSI's imposed water allocation rules, have been a conflicting issue between the DSI and the agricultural agency, which has not been politically clarified. The DSI has had no incentive to make changes, because demand for reduced water inputs was weak and change would have been costly, both for the DSI and the farmers.

* Since all available water could be used for the areas of the three stages, there has been no incentive for the DSI to effectively utilize this potential, all the more so as the water flows at night are not fully exploited. The water available suffices to supply the entire service area except in very dry years. The DSI actually considers that, with the extension of the DSI's service area in the stage IV project area, a higher water-use efficiency will have to be achieved.
* The adaption to suit local conditions, which could reduce water inputs per hectare/crop, makes cooperation with the agricultural agency necessary, and would have increased the DSI's costs for information and operation.
* It might also be difficult, and costly, to change the rules to which the farmers have become accustomed, and which, until now, have promoted irrigation and not the rationing of water.
* The demand for water savings, as a means of groundwater level and salinity control, from the farmers within the project area has been weak, because within the service area high groundwater levels and saline soils have been reduced with the installation of the main drainage networks.
* Water savings are, however, not an option which would have effectively solved the problem. As was previously mentioned, high groundwater levels have been the main obstacle for cropping throughout the year, because the winter and spring rains are not effectively drained out from the area until April, due to inadequately maintained drains.
* Even in the stage IV project area where groundwater levels and saline/alkaline soils have increased, reduced water inputs in the upstream areas would not have solved the problem because groundwater is to a considerable extent recharged through precipitation. The solution to the problem in the stage IV project area is the extension of the drainage networks.
* To date, water demands from other sectors, which could have instigated water savings in this agricultural area, have not been a matter of importance: an expansion of energy generation is not considered by the power plant operating unit; the domestic water supply for rural and urban areas is satisfied by other surface and groundwater resources which have water with a higher quality; the 'peace pipeline', which will extract water from the Seyhan and Ceyhan rivers and deliver it to the Middle East countries, is, at present, an unrealized option.

Taking all this into consideration, the DSI's logic of action followed a cost-minimizing principle, and it has been under no pressure neither from within the project, nor other sectors, to save water

2. The DSI, and particularly the Water User Groups could not successfully enforce the water supply schedules.

The water supply schedules have been poorly enforced by the Water User Groups' ditch-riders, and water was applied in excess. The extent to which farmers apply more water than assigned by opposing the water supply scheduling cannot be estimated. It has been, however, repeatedly reported by the DSI's staff and the WUGs that the farmers apply too much water, exceeding the assigned amounts. On-farm irrigation efficiencies have been estimated for various crops (e.g., cotton, maize, citrus, soybean) comparing the applied amount of water with the amount of water consumptively used. A mean application efficiency of 50% is reported, i.e., 50% of the applied water is lost to runoff and percolation, the latter producing high groundwater levels. Reasons for low application efficiencies are irrigation practices, irrigation technology, planting practices and land levelling. In 1989/90, for example, more water had to be released, because on 31,819 ha corn was not planted in rows. In cotton farming, the irrigation works are done by temporary employed labourers who are not familiar with proper irrigation practices. Because they are paid per hectare, they try to shorten the irrigation time by using more siphons than permitted, flooding the fields. Or, they probably try to compensate poorly levelled fields by applying more water in order that the water may reach the down-lying parts of the fields. These practices increase local surface runoff and percolation, and, by overtaxing the supply schedule, an artificial water shortage is produced for the downstream irrigators. The stricter enforcement of the water supply schedule by the Water User Groups could influence irrigation practices but monitoring costs are high. The consequences remain, however, excess water inputs. The reasons which could explain the demise, are the role of ditch-riders; the weak and conflicting position of the headmen, i.e., to act as service suppliers and conflict mediators, and as politically elected mayors, and the fact that sanctions are, however, weak and rarely enforced.

Two factors have generated uncertainty, and incentives for the farmers not to follow the supply schedule. The first factor refers to the definition of user rights in times of scarcity when some farmers are deprived of their claims, and the second, which is actually more important, relates to situations of peak demand, when the capacity of the delivery channels does not suffice to satisfy the demands of all farmers, particularly, when all farmers have planted high water-consuming crops. A proportional reduction of water deliveries would be a possible solution as the risks would be spread and shared amongst the beneficiaries. The implementation would increase cooperation and operation costs, however, if compared with the present rule.

3. The DSI was not able to prevent irrigation outside its service area, even with saline water.

The extent of affected lands would have remained limited if the state agency had been able to restrict access to irrigation and drainage water usage in the not yet connected stage IV project area. The DSI's staff, however, has not been able to do this because of high monitoring and enforcement costs, (e.g., there are no Water User Groups,) and thus has legalized this practice by charging for the water. But even in this area, the main solution to the problem is the extension of the main drainage networks and the construction of the drainage pumping stations, with investments being required.

4. The DSI has regarded maintenance as the most effective groundwater table control measure.

In the DSI's annual inspection reports on groundwater and salinity, the causes of rising groundwater levels and recommendations for their alleviation are documented, and they are, as far as they concern the DSI's activities, transformed into annual maintenance programmes. Since the Water User Groups have participated in annual field inspections, information on local maintenance and repair needs has been available, and the farmers themselves inform the DSI's staff if delivery channels are broken or in disrepair. The state agency could not uphold the maintenance activities for the deliveries resulting in unabated seepage, even when they received major attention in the programmes. From the data available it can be concluded that the DSI's maintenance activities, especially the maintenance works on the drains, were seriously hampered and the quality of services provided devaluated. In this field the DSI's dependency on central budget allocations, and on its shortages, has shown the most negative effects.

5. The assessment procedure for O&M water charges was deficient, and the collection procedure encouraged free-riding.

Water charges had never played a regulatory function with regard to the rationing of water, and their assessment had almost no effects on water use at on-farm level, e.g., irrigation and planting practices, and on land levelling. But even cost revocery could not be achieved due to the ex-post facto assessment procedure; their poor collection was the cause that the beneficiaries' contributions were very low, and that the provision of irrigation services was almost totally subsidized, as the O&M budgets were set regardless of the low repayment rates. In 1983, the subsidies for irrigation services were 87.3%, and in 1991, 82.8%. No political decision has effectively addressed the high rate of non-payers and this particular form of indirect subsidy. This characterizes the relationship between the government and the beneficiaries, who form the political parties' clientel and have strong pressure groups. The policies of the pre-1980 governments were characterized by import subsidies, low interest credits, tax exemptions, the provision of irrigation projects, and tariff protection. The farmers enjoyed, however, high net incomes, and incomes from agriculture have only been taxed since the middle of the 1980s. Subsidies have been provided for water, fertilizer, pesticides, output prices, credit terms, and the provision of irrigation infrastructure facilities. Between 1960 and

1980, the farmers' lobbies, in particular, the Turkish Farmers' Federation, successfully opposed all attempts to increase the financial burden on the farmers (e.g., the extension of income taxation to include the agricultural sector, and the imposition of other taxes). The farmers' lobbies are the Turkish Union of Agricultural Experts, the Turkish Agricultural Chambers, and the Turkish Farmers' Federation. The latter is a voluntary organization set up by farmers from the Mediterranean and the Aegean regions. The prosperous cotton farmers are the main force behind the Federation.[78] The situation changed significantly, however, during the 1980s, when taxes on agricultural incomes and on various crops were levied, and subsidies on inputs were removed. These were the main areas in which the government changed its agricultural policy.

The cost recovery issue has been set on the agenda; but, at that time, it was believed that cost recovery could not be achieved unless the operation and maintenance responsibilities were transferred to Water User Associations. The exclusion of free-riders was politically not desired, e.g., it would have been costly in an administrative and political sense, and the collection problem could not be solved.

2.
The Second Public-Farmer Setting: Joint Operation and Maintenance by the Regional Directorate for State Hydraulic Works and the Water User Associations

In 1986, a rehabilitation programme for the stage I,II and III project areas was negotiated between the World Bank and the Turkish Government, and a commitment was set by the World Bank that capital and recurrent cost recovery had to be achieved by the beneficiaries. The DSI stressed the point that cost recovery would be impossible to attain under the prominent institutional setting, and proposed that operation and maintenance responsibilities in the DSI-managed large-scale irrigation projects should be, whenever possible, transferred to water user organizations. This initiative was encouraged by the World Bank's policy.[79] In 1993, the accelerated irrigation transfer programme started, beside others, in the Lower Seyhan Irrigation Project.

---

[78] See Ergüder 1991

[79] Plusquellec (1995) describes the World Bank's role: "One of the major contributing and motivating factors for accelerating tranfer programmes in Turkey was the exposure of the DSI's staff to the transfer of irrigation districts in Mexico. After participating in the May 1993 workshop for Bank's staff on irrigation management transfer in Mexico, Joma Mohamadi from ECIAE, persuaded DSI to accelerate transfer of irrigation systems and to organize visits to Mexico. Five groups totalling about 60 high-level officials from DSI headquarters and regional offices visited transferred irrigation districts in Northwest and Central Mexico in 1993 and 1994. This visit had a substantial effect in further encouraging DSI staff to pursue an accelerated programme." (p.1-2)

2.1
The DSI's Initiative to Transfer Large-Scale Public Irrigation Schemes to Water User Associations

From the 1950s until 1993, the General Directorate for State Hydraulic Works only transferred small and isolated irrigation schemes which were difficult and uneconomic for the DSI to operate.[80] The establishment of Water User Groups at the lower levels of large-scale systems is viewed as a positive step towards ensuring participation and developing a sense of cooperation and organized management skills. The DSI still considers them „to be highly appropriate transitional (intermediate) organizations for the gradual establishment of successful Water User Organizations".[81] Water User Groups will be introduced in the stage IV project area, and in the large irrigation schemes of the Southeastern Anatolian Project (GAP).[82] Since 1993, the DSI has initiated the transfer of large-scale public irrigation schemes to varying types of Water User Organizations because „the O&M financial burden for DSI and the Government was getting unbearable and unsustainable. The O&M cost recovery, largely due to political reasons, has been unsatisfactory (about 30%). Considerable increase in the costs of O&M due to the role of unionized labour further aggravated the situation. The present Government's general policy of promoting privatization, (and) positive results from generally satisfactory O&M of transferred schemes, was another important contributing factor, which substantially alleviated the concern that the systems will rapidly deteriorate after transfer".[83] Continous budget shortages for O&M works resulted in the deferred maintenance and poor operation of the projects. The DSI outlines the difficulties with the operation and maintenance in large-scale irrigation systems:

> Neither DSI nor the farmers feel a sense of ownership for the irrigation infrastructure. Through a tradition of paternalism, the farmer has developed an unreasonable over-dependence on the government to coordinate everything. This attitude increases the DSI's burden of physical maintenance of the schemes.
>
> The DSI's practice of carrying out all its routine annual maintenance which consists mostly of maintenance and repair of canals and canalettes, sediment removal and weed control along waterways, grading roads, has been increasing the financial burden of O&M costs which is becoming unbearable for the national budget.
>
> In an average state-managed irrigation scheme, 73% of the O&M costs goes towards operation costs covering personnel, energy, fuel and transportation.

---

[80] Within the territorial boundaries of the DSI's Sixth Regional Directorate, the Bozyazi Water User Association has an irrigable area of 1,370 ha, the Akdeniz WUA commands 1,000 ha lands, and the Limonlu WUA 63 ha.

[81] Mohamadi and Uskay 1994, p.10

[82] Management sub-models are evaluated for the GAP irrigations projects. Although the main system could be managed either by an autonomous Irrigation Agency or a Large Private Company, the models having fully involved Water User Groups on tertiary levels received the highest assessments, regardless of the main system's ownership and management structure. (see Ünver et al. 1994)

[83] Mohamadi and Uskay 1994, p.6

The DSI is increasing the irrigation area annually by 60,000 - 70,000 ha. This puts the O&M Department of the DSI under pressure to keep a satisfactory level of O&M service in the future years with declining personnel, equipment and budget.

There is a 10% penalty for late payment applicable once to late payment of water charges which in an environment of high inflation encourages late payment. This results in poor collection. A new penalty provision for late payment is under consideration by the government to improve the collection of water charges.[84]

However, the political emphasis on the transfer of O&M responsibility to Water User Associations primarily stems from the political wish to lift the financial burden from the national budget. It has been estimated that the transfer would reduce the annual O&M expenditures per hectare by about US$ 100, and that the Turkish Government could save about US$ 10 to 16 million each year.[85] In addition, there were also national restrictions on agency growth, and there was a complete stop of new assignments to the O&M departments.[86]

2.2
Transfer Experiences in Publicly Financed Minor Irrigation Schemes

In Turkey, irrigation systems can be legally transferred to various types of water user organizations such as Water User Associations (WUA), Municipal Organizations, Village Organizations, and Cooperatives (see Box 4.2.).

The General Directorate for Rural Services is entrusted with the construction of small surface irrigation schemes (< 500 l/s), and it has experienced and established methods of transferring O&M responsibility to water user organizations and to local governments, because it has not been able to execute operation and maintenance activities in thousands of facilities scattered all over the country.[87] Within the boundaries of the GDRS's regional directorate in Adana, 21 schemes have been transferred with 9,108 ha and 3,389 beneficiaries. Before 1992, the GDRS's surface irrigation schemes were turned over to user cooperatives in an informal manner, because, until 1992, there was no legal requirement to form cooperatives for surface water schemes. The responsibility for O&M was taken over by the village headmen. When surface water schemes linked several villages, however, conflicts occurred between the villages. The major unresolved problems were the determination of the water rotation order, and the collection of water charges, and many surface schemes were, therefore, abandoned. After 1992, the formation of cooperatives has become a prerequisite for all initiated projects, similar to the establishment of groundwater cooperatives (see Annex 11).

Before 1993, only small and isolated surface irrigation schemes had been transferred from the DSI to Water User Associations. Eight Water User Associations exist in the provinces of Adana and Icel.[88] The DSI remains the owner of the fa-

---

[84] Mohamadi and Uskay 1994, pp.8-9
[85] Ibid., p.4
[86] See DSI, IIIMI and EDI 1996, pp.16-17
[87] Sayin et al. 1993, p.4
[88] Bozyazi, Akdeniz, Erdemli, Limonlu, Hacibeyli, Sumbar, Çukurova University, and Ismetpasa Mahalli WUA (Tekinel et al. 1993, p.12)

cilities, granting user rights to the WUAs; the WUAs have to pay the initial investment costs, and are fully responsible for the operation and maintenance of the entire system at their own costs. If a WUA is not able to maintain the scheme with their own forces, the DSI provides services which are charged to the WUA. The WUA determines the water allocation order, and the assessment and the collection of water charges. The major difference, when compared with the newly established WUAs, is that they are smaller in size, and to a larger extent autonomous in ruling their own affairs. Another difference refers to the representative principle: the number of elected delegates of the (old) WUA parliament depends on the area of irrigable land belonging to a village, and not, as it is with the new WUA, by the area of land of the landowners who joined the WUA as registered members. In the old WUAs, the delegates are elected by the village assembly, while the delegates of the new WUAs are elected by the WUA members.

Both state agencies, the DSI and the GDRS, agree that, under certain conditions, water user organizations „have generally demonstrated the ability to operate and maintain the systems satisfactorily through recruiting required staff, buying urgently needed transport and communication equipment, assessing and collecting water fees, equipping their offices and substantially improving water delivery at a cost generally less than that incurred by DSI".[89]

Box 4.2. Types of Water User Organizations[90]

First Type: Water User Associations
An irrigation scheme can be transferred to a WUA where there is more than one local administrative unit (village legal entities, municipalities) within one irrigation scheme. These WUAs are established under a statute which has to be approved by the Council of Ministers. (Municipal Law No.1580)

Second Type: Municipal Organizations
A scheme which serves only a single municipal unit is transferred to a Municipal Organization. In this organization, the mayor is the natural chairman, and the agreement of transfer is signed by the DSI and the mayor, proposed to the Ministry of Public Works and Settlement for approval. (Municipal Law No.1580)

Third Type: Village Organization
This scheme serves only a single village. The Village Head (Muhtar) is the natural chairman, and the transfer is signed by the DSI and the mayor, proposed to the Ministry of Public Works and Settlement for approval. (Village Law No.442)

Fourth Type: Cooperatives
The cooperatives are established under the Cooperatives Law (No.1163), and it is mandatory for a legal cooperative to be formed at the request of a minimum of 15 farmers before a new scheme is undertaken.

---

[89] Mohamadi and Uskay 1994, p.4; see Kiratlioglu 1994
[90] Source: Mohamadi and Uskay 1994, pp.9-10

2.3
The Water User Associations in the Project Area

The formation of Water User Associations started in 1994, and in May 1995, all except one of the WUAs were established. (see Annex 8). The DSI's management, the regional managers, and some O&M field staff played a highly initiative role. There was direct interaction between the DSI's engineers and the local authorities, including the provincial governors, the village and municipality councils and chairmen, and leading farmers.[91] The farmers and the local authorities have shown a great receptivity, and all villages decided to join the WUAs.

The service areas of the established WUAs cover the entire areas of the stage I, II and III projects.[92] One WUA's service area is defined around a hydrological unit with a main canal as its centrepiece. The number of villages and municipalities located within a WUA's area of responsibility, the number of beneficiaries, and the extent of the service area differ substantially. During the establishment process, the delegates insisted that their area of responsibility should be manageable, and the size of irrigable land should be sufficient to constitute an adequate budget. The largest WUA commands 16,980 ha with 1,812 beneficiaries, other WUAs have between 12,000 and 7,000 ha, or less; the smallest WUA supplies 1,765 ha. (see Annex 8) In the initial process, two delegates had been elected by each village council, and the village headmen and mayors of the municipalities joined the first assembly as 'natural' delegates.[93]

A two-thirds majority of delegates had to confirm the establishment of a WUA, and it was then legalized by the Council of Ministers. In the General Assembly, the villages are represented by their 'natural' delegates, and the other delegates have been directly elected by the beneficiaries whose number depends on the size of the irrigable land of the registered WUA members.[94] The representative principle renders it necessary that important issues are discussed at on-village level, giving the delegates a mandate for voting in the General Assembly. The parliament of the WUA elects the Board of Directors and the chairman; he is elected for five years, the other four Board members for two years only. The manager (i.e., General Secretary), who has to be an agricultural engineer, the accountant and the irrigation-fee collectors are appointed by the Board of Directors, and they have to be confirmed by the Province Governor. (see Annex 9)

---

[91] See Plusquellec 1995

[92] Within the stage IV project area, Water User Groups will be established.

[93] The village headman, the village council, and the mayors of the municipalities are directly elected by the voters. A municipality is legally established in villages with more than 2,000 inhabitants; villages have less than 2,000 inhabitants.

[94] Prior to 1993, the number of delegates depended on the size of irrigable lands within the village borders.

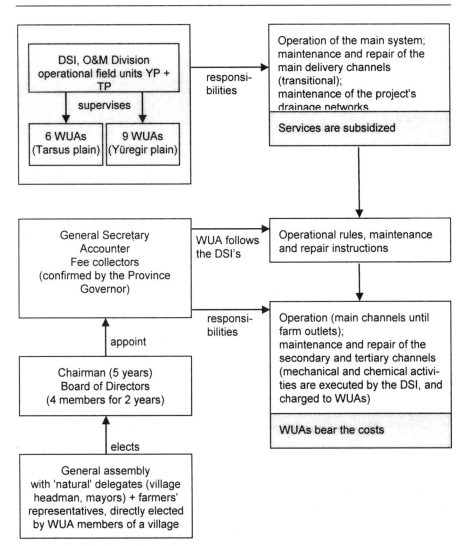

Fig. 4.4. The WUAs establishment, and the relationship between the state agency and the WUAs

All farmers who want to be supplied with services have to become members, and they are requested to pay a one-time membership fee. Most of the farmers have become members since the non-members have to pay double the rate for the O&M services provided, and this has been, of course, the most important incentive to join the WUAs. Not all farmers have been enthusiastic about the transfer of O&M responsibilities. Many farmers who did not repay the recurrent costs to the Ministry of Finance are now obliged to pay the costs, and are aware that if they do not they must pay penalties, and will receive no irrigation water. The South Yüregir WUA, having the largest service area, reported that it has been difficult to persuade the farmers to become members as they had not paid any fees since the operational beginning of the irrigation project. It is reported that the number of farmers who immediately joined the WUA as members was higher in WUAs having a smaller service area. An evaluation team with representatives from the DSI, the International Irrigation Management Institute and the Economic Development Institute considered that the „farmers' principal concern relate to fee payment (after years of very relaxed treatment of obligations), to the quality of service which can be provided by the smaller (WUA) staff, to the technical competence of the (WUA) staff relative to that of the DSI, and to the accountability and fairness of Association management to farmers' needs and wishes".[95]

Water rights are not transferred, and the hydraulic infrastructures remain the property of the DSI who, therefore, exercises control over the WUAs' operation and maintenance activities. The O&M responsibility for the irrigation networks are turned over by contracts negotiated between the WUAs, on behalf of their members, and the Regional Directorate of the DSI. (see Annex 10) The DSI handed over water supply planning, scheduling, enforcement, monitoring, and conflict settlement, and concentrates, instead, on giving assistance and training to the WUAs. The WUAs are fully responsible for O&M works, i.e., they have to organize and finance the operation and maintenance activites. They issue the actual water requirement plans for the planning of the water supply, and the timing schedule, and are responsible for its enforcement, and for the maintenance of their delivery networks. They have to follow the DSI's O&M instructions and the DSI's method for calculating the crops' water requirements.[96] (see Annex 10, Art.4 and 5) This is regarded as a necessity by the DSI, because the WUAs are not yet experienced, and a time for adjustment and training will be needed to even follow the DSI's instructions. Annual maintenance programmes are set by the WUA, and, as long as they have no equipment of their own, the DSI's equipment and staff is rented out to them. Because of their lack of experience, some WUA leaders have felt overwhelmed by the task of planning the annual maintenance programmes. Other WUAs, in areas which already had well functioning and well equipped Water User Groups, immediately started with maintenance and repair works in their main canal although this has remained the DSI's duty. To be successful with the transfer programme, serious training efforts are necessary after the official transfer,

---

[95] DSI, IIMI and EDI 1996, p.22

[96] The transfer contracts can be terminated by the state agency if the WUAs do not follow the DSI's instructions for O&M.

and at least two more years of regular monthly assistance and inspection are required.[97]

Table 4.10. The DSI's and the WUAs' areas of responsibility

	Water User Associations	**DSI, 0&M Division**
Water supply planning Operation	preseasonal, current from the main canals to the farm outlets (following DSI's operation rules)	main hydraulic structures balancing water supply among WUAs of a water subdistrict
Conflict settlement	among irrigators of one WUA	between WUAs
Maintenance and repair	manual works in secondary and tertiary deliveries; repair	maintenance and repair of the main delivery canals
		mechanical, chemical and biological works (on behalf of a WUA)
		drainage networks
		control of WUAs through annual field inspections
Observation programme on groundwater and salinity		DSI's staff only
Maintenance of on-farm drains	nobody is responsible	

The operational staff, i.e., a team of technical supervisors and ditch-riders, is employed by the WUA; one ditch-rider serves an area of between 1,500 to 2,000 ha, which is more than under the DSI's management, and during the irrigation season additional staff is hired on a temporarily basis to assist the ditch-riders (one for 400 to 800 ha).[98] Motorbikes and cars of the former Water User Groups were taken

---

[97] It has been agreed between the DSI and the World Bank that a long process of establishing and training the members of the Water User Associations is required before they can assume their full responsibilities.

[98] The South Yüregir WUA has employed one water distribution technician per 1,500 hectares; in the area of the Tarsus Ten Villages WUA, 9 technicians serve about 12,000 hectares.

over, and became the WUAs' property. The enforcement of the water supply schedule and the settlement of conflict amongst the irrigators and the member villages rests completely within the WUAs' responsibility. The conflict resolution arenas are the village assemblies, the WUA commission, the General Assembly, and the local court system.

A one-time membership fee constitutes the initial budget. This fee is correlated to the size of a member's irrigable land, i.e., between TL 50,000 and TL 100,000/ha have been levied.[99] The WUAs are responsible for the collection of water charges, and the levying and enforcement of penalties. Non-payment is penalized by a 10 % fine (monthly) on the O&M costs, and they are effectively enforced by threatening the non-payers with sanctions (amongst others with water cut off). The operation and maintenance expenditures for the secondary and tertiary delivery systems are fully financed by the WUAs, while the DSI's remaining O&M activities are subsidized. The O&M costs are assessed on a per hectare/crop basis as they were under the DSI's management, but they are assessed in advance, and have to be paid in two or three instalments: the first instalment (20%) has to be paid at the beginning of the irrigation season, the second (20%) during the season in June/July, and the last (60%) has to be paid after the harvest in November.

For 1994, the first year of operation, data are available on the O&M costs in WUA-operated systems. The costs per unit are cheaper than the unit costs for the DSI's services: in 1994, the South Yüregir WUA assessed TL 1,200,000/ha cotton, the DSI TL 2,590,000/ha. The 1995 WUA budget plans have reserved 30% for personnel expenses; the average personnel costs varied from 42 to 49% of the operation costs when the DSI operated the scheme. (see Table 4.11.)

The total O&M costs in the WUA schemes are only 40.9% of the costs incurred by the DSI; the operation costs are 45.7 and the personnel costs are 31.8%. The personnel costs, which are the main share of operation costs, are lower in the WUA schemes, but the equipment costs are higher. Of particular interest is the employed personnel (per 100 ha), and the associated personnel costs in the different systems, which range at approximately one third of the DSI's costs. The total number of employed personnel, 0.31 per 100 ha, is sufficient if compared with the indicator set by Bos and Nugteren.

---

[99] In 1995, the farmers had to pay US $ 1.25 or 2.5 per hectare.

Table 4.11. O&M costs in the DSI-operated Yüregir Plain and in WUA-operated irrigation schemes in 1994[100]

	DSI-operated scheme 50,014 ha size	**WUA (old)** 1,370 ha size	**WUA (new)** 4,824 ha size
Permanent personnel			
(No.)	168	7	13
(TL/ha)	716,000	280,000	226,000
Temporary personnel			
(No.)	45	-	2
(TL/ha)	60,000	-	21,000
Total			
(No.)	213	7	15
(TL/ha)	776,000	280,000	247,000
Equipment (TL/ha)	43,000	-	90,000
Other costs (TL/ha)	46,000	40,000	63,000
Total operation costs (TL/ha)	875,000	320,000	400,000
Maintenance costs (TL/ha)	150,000	52,000	20,000
Total 0&M costs (TL/ha)	1,025,000	372,000	420,000

The DSI continues to operate the main headworks, (i.e., the regulatory and the main canals), and balances the water supply for the subdistricts. Because there is no institutionalized linkage between the WUAs of one water subdistrict, the DSI coordinates conflicting targets. The maintenance and repair of the main delivery channels is still the responsibility of the DSI's staff, because special knowledge and machinery is required which is not yet available to the WUAs. In the future, the transfer of maintenance responsibilities for the main canals is envisaged, and the WUAs would have to consequently raise the service fees. The responsibility for the maintenance of the drainage systems (main, secondary and tertiary) was not transferred, and it remains, together with the groundwater and salinity observation programme, within the DSI's responsibility. All these services are not charged to the beneficiaries.

---

[100] Source: DSI, General Directorate, O&M Department 1995

Table 4.12. Number of personnel, and personnel costs for the different schemes in 1994[101]

Personnel	Yüregir Plain (DSI)	Kuzey Yüregir WUA	Personnel costs (% of DSI)
permanent (No./100 ha) (TL/ha)	0.34 716,000	0.27 226,000	31.5
temporary (No./100 ha) (TL/ ha)	0.09 60,000	0.04 21,000	35.0
total personnel (No./100 ha) (TL/ ha)	0.43 776,000	0.31 247,000	31.8

2.4
Conclusions on the Second Public-Farmer Setting

Although it is too early to draw valid conclusions, and it is not yet proven if the WUAs contribute towards better performance in the long-term, the transfer of O&M responsibilities to the WUAs provides some advantages if compared with the first public-farmer setting (see Table 4.13.).[102] Above all, the second public-farmer concept clearly defines the responsibilities and the WUAs' areas of decision-making, and it determines which state-provided services are subsidized and for which services charges are levied. The system is operated by the WUAs' staff on a 24-hours basis, and the supply scheduling is controlled both by their staff and the WUA members themselves, having the potential to be more effectively enforced.

The operation of the system has improved, and, as a first review underlines, the irrigation efficiency has increased, for example in the area of the North Yüregir WUA, from 40% under the DSI's governing to 59%, because the WUA fully utilizes the 24-hour water deliveries.[103] After the first irrigation season 1994/95, the farmers expressed their satisfaction with the service provided by the WUAs. In their perception, water supplies were more adequate, reliable and controlled. The most important reason, in their view, is that the employed personnel has worked more efficiently than the DSI's staff, because its working hours are more compatible with the operation of the irrigation scheme, or as a WUA chairman said: „The staff works at a time and in a way we demand it." The costs have to be fully recovered by the beneficiaries. The way in which the water charges are collected

---

[101] Source: DSI, General Directorate, O&M Department 1995; Mohammadi and Uskay 1994

[102] Water User Associations were recently established, and they had experienced only one irrigation season when the interviews were made.

[103] DSI, General Directorate, O&M Department 1995

from the WUA members, i.e., two or three installments, fits in more with the farmers' ability to pay; their payments are strictly controlled, and sanctions on non-payers are levied and executed. Controlling high groundwater levels and salinization remains weak because there are no incentives set for the rationing of water: as the WUAs have to follow the DSI's operation rules, including the calculations for the crops' requirements, the water inputs per hectare and crop will not reduce, and the water charges are still assessed per area and crop. The state agency, legally entrusted with on-farm water use, still has no means of influencing such matters.

Table 4.13. Comparison between the two public-farmer settings

DSI - WUG	DSI - WUA
High water inputs	
per ha/crop as determined by the DSI	per ha/crop as determined by the DSI; better conditions to adjust water inputs to suit local conditions
Water distribution problems and excess water withdrawal	
* low labour productivity of the DSI's staff * little monitoring by the DSI * higher water inputs than assigned * low sanctions for rule opposers	* working hours fit in with need * higher monitoring * scheduling is enforced * water cut-off; fines
Maintenance of drainage systems	
DSI's activities depend on budget allocations, available equipment, labour productivity high rate of non-payers	DSI's activities depend on budget allocations, available equipment, labour productivity subsidized
Maintenance and repair of delivery systems	
DSI's activities depend on budget allocations, available equipment, labour productivity minor contributions by WUGs	* reduced DSI's activities depend on budget allocations, available equipment, labour productivity * WUAs have full responsibility for maintenance of secondary and tertiary systems, can effectively reduce water losses
Financial resources	
(DSI) central budget allocations: low collection rate to MoF (WUG) beneficiaries' contributions: higher collection rate to WUG	(DSI) central budget allocations: subsidized (WUA) paid by beneficiaries: high rate of collection

Chapter Five
Effects of High Groundwater Levels, Waterlogging and Salinity on Farm Economy

The analysis thus far has shown that the major factors which have contributed towards high groundwater levels, waterlogging and salinization, are related to the main system's operation activities and the poor maintenance, particularly, of the drainage networks, and to insufficient public investments in drainage systems. These factors either have pre-determined high water inputs at on-farm level, or have been the cause of agricultural effluents not being drained out of the area. In the analyzed institutional settings, i.e., the two public-farmer models, the farmers have no control over investment decisions, and the village-based Water User Groups had no influence on central management issues as they only technically assisted the state agency in operation and maintenance. The implementation of some groundwater and salinity control means are, however, within the scope of the farmers' decision-making: the farmers could invest in supplementary on-farm drainage systems as it was expected by the state, and they could maintain the state-installed and owned on-farm drains, but they did not. The farmers control the use of water at on-farm level which shows poor application efficiencies due to poorly maintained land levelling, and improper irrigation and planting practices. As water charges have been assessed at levels to promote irrigation and the planting of specific crops, they could not function as an allocative device to save water. It is assumed here that the farmers respond in a rational economic fashion when they use 'too much' irrigation water, and when they substitute water for higher priced inputs such as labour and technology, because the water charges are area-crop based and not related to the amount of water used, and water savings do not make sense for the farmers.

This chapter evaluates to which extent the farmers are negatively affected by high groundwater levels and salinity, and compares the positive impact of the irrigation project on cotton farming, e.g., the development of yields and incomes, with the negative effects which emerge from groundwater and soil conditions; it concentrates on cotton farming, as cotton had, for a long time, been the dominating crop. It analyzes whether the farmers have a vital interest in taking preventive and corrective measures to halt and reverse the negative process. It is assumed here that if a farmer's benefit deriving from control means exceeds the costs for water-saving and drainage means, control measures are likely to be implemented. Controlling groundwater and salinization by whatever means poses particular problems, and collective action is regarded to be a conditio-sine-qua-non because if only a few farmers in a drainage sub-basin save water, the impact on groundwater levels is neglible. Investments in drainage systems, and their maintenance, exceed a farmer's capabilities, except when it is a matter of the on-farm drainage systems. But solutions to this particular issue have faced great difficulties. The conditions under which collective action is forthcoming are decisive, and depend on incentives set, and on rule enforcement.

1.
Agriculture in the Lower Seyhan Plain Prior to the Lower Seyhan Irrigation Project

Until the middle of the 19th century the Yüregir and Tarsus Plains were of little agricultural importance. From the 110,000 ha which were agriculturally utilizable in the Yüregir Plain, 6,250 ha were used to cultivate grain, cotton, and sesame; areas further away from settlements were used to grow rice, and areas near to towns and villages were laid out as gardens by Fellaches, who were expelled from Egypt in the first half of the 19th century.[1] The Yüregir Plain was mainly wooded with oak, which spread from the Southern district town of Karatas to Adana, while marshland and steppe dominated the Tarsus Plain. The Yüregir Plain was the winter grazing area of the normadic tribes. It is documented in old travel accounts that the plains were in the hands of tribal and nomadic leaders, who, towards the end of the 19th century, transformed it into independant territories and into land which was regularly used as transit and pasture land. Permanent settlements could only, therefore, develop outside this area.

In 1858 a new land legislation was introduced which brought radical changes to the development of settlements and agricultural use. Until 1858 all state-owned lands were managed by the farmers as hereditary leaseholders. The most significant change was the introduction of a legal title that was almost equivalent to a title of ownership. Up to this point in time, there was only the traditional right to own house plots. The farm land became land with a possessory title for which tax had to be paid, but which could now also be sold, bequeathed or mortgaged. The title was almost in accordance with that of ownership, with permanent cultivation forming the only condition for transfer.[2] The new land legislation was the deciding incentive for the indigenous nomads, i.e., the Yürük and Turkmen, to settle, and for the settlement of Turkish and Kurdish immigrants from Eastern Anatolia, who established separate villages. Religious refugees (the Muhacir) from the lost provinces of the Ottoman Empire, Muslims from Southern Russia (the Nogai-Tartar) and from the north-western Kaukasus (the Turkish Kaukasians and Tscherkessians) settled in the Çukurova, followed later by immigrants from the Balkans. In the 20th century, single Yürük tribes settled in areas still available in the marginal districts of the outer delta regions, their former winter pasture areas.[3] The civil

---

[1] The history of the settlements in the Yüregir Plain is verified by the work of Mustafa Soysal (1976); no comparable studies are available for the Tarsus Plain.
see Kotschy 1858; Schaffer 1903

[2] Hütteroth 1982, p.228

[3] Little new settlements have been established in the Yüregir Plain since 1900. Some villages in the floodplains of the Seyhan Dam were re-settled on the plain. Four vil-

legislation for the equality of the Christian minorities during the Tanzimat period of reforms made it also possible for Greeks and Armenians to make use of the land legislation. The end of the Ottoman rule, the expulsion of the Armenian population, and then of the Greeks after World War I and the period of the War of Independence, „all greatly upset the land tenure situation in this region. In 1923, after the War of Independence, there were extensive areas on which no property rights could be claimed. But as land was plentiful and title deeds not a matter of importance up to the end of the 1920s, squatters cultivated the land undisturbed".[4] A struggle for land in the region, however, resulted in the emergence of a new group of landlords who allowed the squatters to remain as share-croppers in exchange for half of the annual crop. The changes were not related to the system of large-scale landlordism, which had existed before, but did pertain to the persona of the landlord, which changed.[5] They invested a part of their capital in more agricultural land during the 1930s and the early 1940s.

The purchase of land by Armenians and Greeks, and also by Turks, led to the formation of the so-called 'Çiftliks', which are a prominent feature in a number of villages. Çiftlik is the name given to a large private or state-owned farming enterprise in which land accumulation has resulted from prudent management of existing land, labour, and capital, rather than by inheritance. Their landowners have continued to play an active role in farm operation and management rather than becoming absentee landlords. The owner-operators of the Çiftliks had been, in most cases, villagers; sometimes their owners are a community of heirs, whose farms are operated by managers. More than 90% of the Çiftliks in the Adana Plain are smaller than 660 ha, although there are 6 which are larger than 1,000 ha. For the most part, these holdings are not fragmented, as the many villager holdings are. The Çiftliks are concentrated south and west of the city of Tarsus, in the low-lands south of the Tarsus-Adana highway between the Seyhan and Berdan River, and south of the city of Adana between the Rivers Seyhan and Ceyhan. The larger Çiftliks are located in the reclaimed areas along the Seyhan and Ceyhan rivers, and the smaller ones are concentrated along the major transport roads.

Several factors have encouraged the development of commercialized agriculture, of which the most important are the development of transport systems, i.e., the accessibility to the markets; the opportunity for the accumulation of capital, because the owner-operators were wealthy city merchants who had accumulated large capital reserves in trade and finance, and who controlled marketing facilities for agricultural produce; the growing markets for crops and secondary demand for agricultural produce by several manufacturing plants (cottonseed for oil and seedcake; yarn and textile). The Çiftliks concentrated on the cultivation of cotton, ses-

---

lages, situated on the banks of the Seyhan, were moved behind the dykes. Agricultural workers without property were allocated an area of 500 square meters. Residents with farmyards received an area of corresponding size (800-1000 square meters). The allocation of these plots compensated for the lost land. The Government paid a grant to each family for a newly built house; the amount granted was based on the value of their property in the old village.

[4] Kiray and Hinderink 1968, p.499
[5] See Kiray and Hinderink 1968, p.500

ame and rice, which generally command a higher return on the commercial market, and they invested in the agricultural as well as in the urban sector. Some owners of Çiftliks became city merchants, and others invested substantial capital in cotton gins, small textile plants and grain mills.[6]

A great chance was brought about by the introduction of a new cotton variety in 1943, the medium staple 'Akala', which increased yields of seed cotton from 500 to 1,000 kilograms per hectare. Up to 1940, short staple cotton was grown in rotation with wheat; in some areas rice plantation dominated. Fertilizers and pesticides were not used, draught animals were employed for ploughing, and only wooden implements were available for tilling. With the new cotton variety, the value of land increased in relation to labour costs due to higher yields and better prices for the 'Akala' cotton. Of more importance, however, was the increase in the area planted with cotton, because cotton was more profitable than cereal. Wheat growing disappeared completely, and in some areas crop rotation was abandoned. Share-cropping arrangements underwent a change; they took on a businesslike character, and they were no longer limited to the family members. The fifty-fifty share-cropping arrangements were replaced by one-third or even one-quarter systems where the landlords claimed two-thirds or three-quarters. With the disappearance of crop rotation, the contracts were only binding for one season instead of an indefinite period, or two years, as it was before. The monocultural cropping pattern heavily influenced the labour pattern; instead of a fairly equal distribution of farming operations over the year, as was the case in cotton-cereal farming, cotton monoculture reduced labour requirements to only five months, and caused seasonal unemployment for the rest of the year. With 'Akala' and other new varieties, like 'Coker' and 'Deltapine', a labour shortage occurred because the bolls of these varieties open in September and have to be picked within a short period in order to prevent damage by the early autumn rains. Seasonal workers from outside the villages and from outside the region were hired, and their numbers reached a maximum of about 300,000 annually.[7]

In the 1950s, taxes on agricultural products were abolished, and credits were given for the modernization of the farms and for the improvement of production methods.[8] The heavy mechanization, in particular, completely changed the farming pattern, and in some areas, share-cropping gradually disappeared and renting expanded. But renting additional land is only common for large holdings. Scholarly studies question whether mechanization led to unemployment and rural migration during the 1950s. Cultivable land was still available, so even if tractors replaced share-croppers on large estates, they could acquire new land, or find new jobs.[9] Kiray and Hinderink emphasize that mass mechanization contributed to the disappearance of share-cropping, because it was possible for one tractor and its equip-

---

[6] See Hilton 1960

[7] See Kiray and Hinderink 1968; Soysal 1985

[8] The agricultural development in the Çukurova was certainly also influenced by the eradication of malaria, which had been brought under control in the middle of the 1950s.

[9] Kasnakoglu et al. 1990, p.107

ment to displace ten farmers, together with their animals and most of their wooden implements. This trend was observed in villages in the Yüregir Plain.[10]

However, structural changes in the agricultural sector concerning the land ownership pattern, together with the changing of share-cropping arrangements, the mechanization of farm works and the prominent cropping pattern had occurred before the Lower Seyhan Irrigation Project began. While production increases can be attributed to the expansion of the cultivated area until World War II, improved productivity occurred, essentially, due to new seeds, improved farming practices and the installation of the irrigation systems in the 1940s. The favourable conditions stimulated sector developments like second markets for agricultural produce, marketing and transport facilities, and governmental agricultural policies, e.g., input and output price supports. Since the 1950s, „almost all major agricultural products are under government support schemes, constituting over 90% of the total value of agricultural production, and the state institutions purchase a substantial portion of marketed production. Similarly, most modern inputs are either produced, distributed, or priced by the government. Infrastructural investments in roads, irrigation schemes, land development and conservation, and extension services also provide inputs to agriculture free of charge or at highly subsidized prices".[11]

The Agricultural Census in 1963 recorded a highly concentrated land ownership pattern, in comparison to other regions in Turkey, with severe inequality of land holdings among the farmer families.[12] As its data indicated, 33% of the farming families owned no land, while 23% owned 80% of the land; the last category comprised a high percentage (19%) of absentee landlords owning 53% of all lands. About 30% of the farm families did not own land of sufficient size, and agriculture was still at a subsistence level. The many smallholders were regarded as a potential which the irrigation project might force away from the land.

The high degree of landless farmers and the high percentage of absentee landlords were subjects of dispute. Usually, the category 'absentee landowner' indicates that these landowners are not the farm operators, and that, deriving only a part of their incomes from farming, they have little interest in investments in the agricultural sector, but in consumptive uses, and/or in investments in the urban sector. This assumption did not hold true for the 'absentees' in the Lower Seyhan Plain: the majority of absentee landlords operated, and continue to operate their own farms, although they are living in the cities. With its improved roads and transport systems, the Seyhan Plain is easily accessible, and the mechanization of farm operations has made urban residence, with its higher living standards, increasingly attractive. Most of the farmers spend much of the year in Adana or Tarsus, because farm operation activities last for only five months of the year since

---

[10] Kiray and Hinderink 1968
[11] Kasnakoglu (1986), quoted from Kasnakoglu et al. 1990, pp.125-126
[12] Agricultural Census 1963
The data have to be interpreted with care, because little land was registered. Between 1951 and 1971, only about 88,000 ha of land were registered as private property with title deeds, some 6,500 ha were owned by the state, and the ownership of about 30,000 ha was disputed, but they were used agriculturally.

the cotton-wheat crop rotation was abandoned. These 'absentee landlords' form a part of the particularly high innovative agricultural structure of the Seyhan Plain. The high degree of landless farmers, as cited in the agricultural census, is regarded as a statistical artifact by some authorities, because this category included a large number of young farmers working on their fathers' holdings, while others, by improving their skills over the years, are tractor drivers, mechanics, dairy workers and livestock breeders.[13]

A much more realistic picture of the landownership pattern in the plain is provided by Bülbül (see Table 5.1.).[14] From 29,211 farm families, 19,876 (or 68%) owned land, and 9,355 families (or 32%) were landless. Small landholdings dominate in numbers; 72.86% of the families owned farms between 1 and 5 ha which provided insufficient incomes; 15.2% of the families owned farms with a size between 5 and 10 ha, and 11.94% owned holdings greater than 10 ha. These holdings were widespread and owned more than 50% of the total land, of which those greater than 20 ha owned one third.

Table 5.1. Landownership patterns in the project area in 1973[15]

Size of holding [ha]	Number of families	[%]	Area [ha]	[%]
0.0- 1.0	4,777	24.03	3,152	2.86
1.1- 2.5	5,247	26.40	9,614	8.73
2.6- 5.0	4,458	22.43	16,696	15.16
5.1- 7.5	1,848	9.30	11,506	10.45
7.6- 10.0	1,172	5.90	10,601	9.63
10.1- 20.0	1,503	7.56	21,856	19.85
>20.1	871	4.38	36,697	33.32
total	19,876	100.00	110,125	100.00

The land ownership pattern in the Yürük and the Sarkli villages showed substantial differences.[16] In the Yürük villages, there was a lack of large-scale landowners and there are few landless farming families, while in the Sarkli villages a high % of landless families prevailed. An explanation for the differing landowning structures is offered by the historical development of the settlements. The settlement of the Yürük took place in groups, or family groups. Each family within a village or settlement was allocated land by the tribal leader in accordance to the degree of relationship. In the time following this, the villages did not accept any more immigrants. The lack of large-scale land-holdings originated from the situation at the

---

[13]See The World Bank 1985
[14] Bülbül 1973, pp.67-68
[15] Source: Bülbül 1973, p.67
[16] The Yürük were indigenous nomads, and the Sarkli were Turkish and Kurdish immigrants from Eastern Anatolia.

time of establishing the settlement, when the tribal leader did not hold such a dominating social position any more, and only small areas and bad ground were available. Additionally, the right of inheritance stipulates the division of the land (principle of land division), leading to a constant carving up of the land. The high number of landless families in the Sarkli villages, unlike in the Yürük villages, can be traced back to the further immigration and settlement of landless families from Eastern Anatolia, who found employment as agricultural and seasonal workers there.[17] The introduction of the new cotton varieties and the mechanization of farm works instigated changes also occurred in the Yürük villages. There, however, most villagers were in a position to secure their own food through growing cereals and tending home garden plots. The nuclear families of father and sons have become involved in a type of reciprocal relationship. In their separate households, they have worked out share-cropping arrangements with their parents. The sons provide labour, and the crop is shared according to need, meaning that there are no rigid proportions agreed upon in advance.[18]

When Bülbül's figure for landless families is compared with that of the Agricultural Census, it is slightly lower. He differenciates among the landless families between labourers, share-croppers ('Ortakcilik'), and renters of land ('Kiracilik', or time-leasing) which have been two important types of leasing systems existing in the plain. (see Table 5.2.) With time-lease systems, the owner lets the land to the leaseholder for a limited period of time, and at a fixed rate, which depends on the quality of the land and on irrigation possibilities. The costs for operation and equipment are borne by the leaseholder; contracts are concluded for a period of one or two years. With share-cropping, the owner provides the land and contributes half of the running costs.

Table 5.2. Landless families and leasing systems in the project area (in 1973)[19]

	Number	Percent of landless farming families
Farm families (total)	29,211	-
Land owning farm families	19,876	-
Landless farm families	9,335	100.00
Share-croppers	1,234	13.22
Leasers	36	0.39
Land labourers	8,065	86.39

---

[17] Soysal 1976
[18] See Kiray and Hinderink (1968, pp.497-527), in particular, the study on the village Karacaören
[19] Source: Bülbül 1973, p.68

The net yield is shared fifty:fifty, or thirty:seventy. No detailed data are available about the distribution of share-cropping and time-leasing throughout the project area. Soysal reports that, in the 1960s, in 10 villages with 434 surveyed farms, 50 were operated under the time-lease system and 48 under share-cropping, while the rest were self-managed. The leaseholders normally come from the village of the landowner; in some villages the majority are relatives of the landowner. Few villages lease land in the districts of other villages. Usually, share-cropping was practised between relatives, but today it is no longer limited to the members of a family.

River adjustment, mechanization, improvement of the cotton seeds, expansion of the transport network and irrigation led to the tremendous increase in the value of land. The prices for agriculturally utilizable land depend on the soil quality and the availability of irrigation; the proximity to the city of Adana also has a great influence on prices. In the 1960s, for one hectare of irrigated land one might have obtained TL 1,500,000 to TL 2,000,000 (before TL 150,000), and for one hectare of dry land the prices increased from TL 100,000 to TL 550,000. There is general agreement that little land is available for purchase, except near the new industrial sites and the city of Adana. Soysal assumes that no land sales can be found in areas far from towns, not even by indebted farmers.[20] They would rather lease their land hoping to self-manage their land again after clearing their debts. As the income from the leasing is high farmers are not immediately forced to sell their land. The property movements in these areas are very low. Many medium and small sized farms also lease out land under time-share due to the lack of personal capital and the unfavourable credit conditions.[21] As the interest rates of the private credit market are high the leasing out of land is considered an appropriate strategy. Other farmers who work in a town can lease out their land under the share-cropping scheme because they are not dependant on the income from agriculture. To date, no open market in land has developed in areas away from the city of Adana, and only small parcels of one or two ha change hands amongst villagers, although the land values have increased.[22]

Meanwhile, between 1960 and 1985, due to the inheritance law, the number of farms smaller than 10 ha, and in particular those smaller than 2.5 ha, has increased, while the number of farms owning more than 20 ha has decreased; 30% of land is owned by farmers with farms of 20 ha and more (4% of all farms), and those smaller than 5 ha own 29% (73% of all farms). The Regional Directorate for State Hydraulic Works and other state organizations estimate that more than two thirds of all farms are operated by their owners, about 12% are rented out, and 17% operate under share-cropping conditions.[23]

---

[20] Soysal 1976, p.121

[21] The Agricultural Bank only grants credits of TL 15,000 for 10 ha of productive land which has already running costs of TL 55,000/ha.

[22] This trend continues until today.

[23] DSI, 6.Bölge Müdürlügü 1986/87; FAO 1988

2.
Positive Impacts of the Irrigation Project on Yields and Net Incomes

By 1949, the area devoted to cotton began to expand rapidly with the replacement of the native varieties by the more stable type 'Acala'. The output increased due to the expansion of the area cropped, and, at that time, the gross income per hectare of cotton was estimated to be almost three times higher than that of cereals. In 1961, prior to the irrigation project, 75% of the arable area was planted with cotton in dryfarming, and, in some parts of the project area, cotton and other crops were irrigated by withdrawing water from the already existing infrastructure and from the rivers.[24] The World Bank estimated that a result of the project would be a more intensive pattern of summer crops, and that the area of cotton would decline to 35%. The areas under fruits, vegetables, and oilseeds would increase, and fodder crops and legumes would be new crops needed in the rotation for intensifying livestock production and for improving soil fertility. It was further expected that about 45% of the area would be used for producing winter crops (,i.e., legumes, winter vegetables, melons, watermelones, and cabbage) without irrigation.[25] (see Fig. 5.1.)

These expectations were not met because of the farmers' decisions. From 1964 to 1981 cotton remained the dominant crop in irrigated agriculture (94 and 77% respectively), and the crop rotation system 'cotton-wheat' was widespread.[26] In the 1980s, cotton was still being planted on more than 50% of the project area, and only since 1989 has the area devoted to cotton declined to the estimated level of approximately 35%.[27]

Prior to the irrigation subprojects, the average cotton yields were about 600 kg/ha (see Table 5.3.). The reasons for the low yields were inadequate moisture during the growing season, the continious monoculture, and the minimal use of fertilizers. The second most important crop was wheat, with a yield of about 1,500 kg/ha. With the irrigation scheme, yields developed much faster than expected by the planners, but the full potential of the area's natural resources, e.g., a diversity of crops like fruits, vegetables, and fodder plants for livestock production, was not accomplished. By 1971 3,800 kg/ha of cotton was obtained compared to the appraised 2,400 kg/ha expected at full development in 1985. The cotton yields were not only substantially higher than the appraised estimates, but compared favourably with yields in other countries.[28] The yields of wheat developed along similar lines; in 1971, due to the introduction of the high yield variety 'Mexican wheat' and the use of fertilizers 3,900 kg/ha were obtained, and an average of 3,600 kg/ha was achieved. The appraisal had estimated 2,800 kg/ha for 1985.

[24] See DSI and TOPRAKSU 1961; Soysal 1976
[25] See IBRD/IDA 1963, Annex 4
[26] Saglamtimur 1988
[27] See Annex 3: Crop Patterns
[28] For example, in Egypt (2,500 kg/ha), Sudan (700 kg/ha), Israel (3,400 kg/ha), Australia (3,600 kg/ha), and in the USA (1,700 kg/ha) (see The World Bank 1985, p.45)

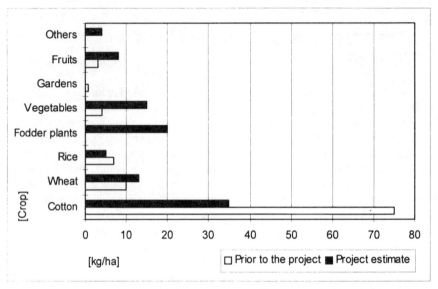

Fig. 5.1. Crop pattern prior to the irrigation project and the projected estimation (area percentage) [29]

The yields of cotton and wheat have fluctuated considerably during the years. Until 1978 yield fluctuations were more pronounced than in more recent years. Average cotton yields from 1970 to 1978 were 3,000 kg/ha, with deviations of up to 30%. Similar developments were observed for wheat, i.e., 3,300 kg/ha, with deviations of up to 33% due to low rainfall in 1972 and 1973. Yield fluctuations reduced until 1984. The largest deviation from the mean for cotton (i.e., 3,500 kg/ha) was 17% which was a result of the higher incidences of insect pests. In 1974 and 1975, the greatest yield depressions occurred. The appearance of the pest 'Bemisia tabaci' was encouraged by the favourable conditions of extremely high temperatures and heavy rainfall (the rainfall in August, September, and October 1974 was twice as much as the average for the preceeding five years). The plants had also been weakened through the use of nitrate fertilizers; these N-fertilizers stipulated the plants to grow quickly, producing an underdeveloped stroma. This increased the plants' susceptibility to the pest, thus enabling it to multiply and spread.[30] The white fly infestation reduced the average net income per hectare by approximately 90%, and therefore the cotton area was drastically reduced.

---

[29] Source: DSI and TOPRAKSU 1961, p.38
[30] See TU Berlin 1992, p.75

Table 5.3. Cotton and wheat yields in the stage I and II project areas from 1967 to 1984[31]

Appraisal estimate	Cotton [kg/ha]	Wheat [kg/ha]
Without project		
rainfed	600	1,500
irrigated	2,000	
With project		
at full development	2,400	2,800
1967	2,200	
1968	2,800	
1969	3,100	2,800
1970	3,500	2,700
1971	3,800	3,900 (high yield variety)
1972	3,300	3,000 (low rainfall)
1973	3,300	2,300 (low rainfall)
1974	2,300 (pest)	3,400
1975	2,100 (pest)	3,500
1976	3,400	3,500
1977	2,600 (pest)	3,500
1978	2,700 (pest)	3,700
1979	3,700	4,100
1980	3,500	3,600
1981	2,900 (pest)	4,500
1982	3,600	4,300
1983	3,600	4,700
1984	3,700	4,100

The land thus shifted out of cotton was largely put under wheat, a risk-minimizing alternative, since it was a well known crop with low production costs which allowed the farmers to overcome monetary constraints. The shift to wheat was also encouraged by the doubling of its support prices from TL 1,200, in 1973, to TL 2,050, in 1974, and to TL 2,400, in 1975. In 1977/78, and in the beginning of the 1980s, pests reappeared, but in the 1980s sufficient effective pesticides had been made available, and, therefore, yield depressions could be restricted.[32]

There have been several governmental attempts to get the farmers interested in diversifying the crop patterns and in second cropping, as the natural conditions in the Lower Seyhan Plain favour cropping throughout the year and a diversity of crops. This would also overcome yield depressions and income losses caused by

---

[31] Source: The World Bank 1985, p.44
The data are only valid for the stage I and II project areas, and therefore do not coincide with the data given by Yurdakul (1986), Yurdakul and Ören (1991) or by Tekinel (1992, p.6).

[32] The most common cotton pests are Bemisia tabaci Genn., Heliothis armigera Hbn., Spodoptera lilloralis Boisd., Agrotis ypsilon Hufn., Tetranychus cinnabarinus Boisd., Platyedra gossypiella Saund., and Earias insulana Boisd. (see Sengonca 1982, pp.51-56)

pest incidences. In 1969/70, the Farmers' Extension Service started its crop diversification effort which concentrated on vegetable production, citrus plantations, and on mixed farming. In the stage II project area, for example, 7,000 ha were suitable for citrus. The farmers showed little interest because they were aware that after planting citrus trees there would be several years of lost income until full production was reached. Unattractive credit terms and badly developed marketing opportunities, together with the fact that the farmers had no experience with these crops, caused them to continue with cotton farming, all the more, the net returns were high. In this case they could fall back on decades of experience, and well developed marketing structures (e.g., a purchasing cooperative and several cotton processing factories are found in the region). In 1970, sorghum was introduced to supply the government mill in Mersin. Sorghum was planted on 400 ha as a second crop after wheat, and achieved high yields. However, there was no subsequent expansion because the government mill halved the price of sorghum just prior to the harvest. Vegetable production was considered risky by the farmers due to the unorganized marketing system, and thus increased little. Only some farms, located near the cities of Adana and Tarsus, started with vegetable plantations on small areas. Because the demand on vegetable fats and oils could not be met by home production, and because their import took up a high share of the total expenditures for imported foodstuffs, soybean and maize have, since 1981/82, been promoted by the provision of credits with attractive terms. The market, however, was not secure. A condition of the purchasing cooperative was that soybeans should have a certain moisture content, which the farmers could not produce due to the lack of processing capacities. Thus the production capacities for soybeans were left unexploited. Only since 1982 has this processing been guaranteed by the purchasing cooperative. When the market had been stabilized for these crops, e.g., output prices, favourable credit terms, etc., the crop rotation system 'wheat-soybean' and 'wheat-maize' increased, and, to a considerable extent, soybean and maize were planted as second crops in May/June after wheat was harvested.[33] As the domestic demand for meat could not be met, interest-free credits were made available to the fatstock farmers. The promotion to intensify livestock production, which depended also on alfalfa, was not successful because credits were only provided to farms with 20 or more cattle. This number was later reduced to 10.[34]

The alternation of the cropping patterns, i.e., the shift away from the dominating crop cotton, has recently become more evident to prevent interregional competition between the Lower Seyhan Plain and the Southeastern Anatolian Project (GAP) region. In the GAP region it is planned to irrigate about 1.6 million ha of land, and cotton should be planted on 25% of the area. The climate in Southeastern Anatolia, with its winter frosts and very hot and dry summers, is a natural constraint to cotton pests, and the high costs for plant protection would reduce. When irrigated agriculture expands, workers will find jobs in the Southeast, and the seasonal migration to the Lower Seyhan Plain of between 200,000 and 300,000 work-

---

[33] Tekinel 1992, pp.7-8
[34] Personal communication Bülbül 1995

ers will come to an end, and thus labour will be a constraint for irrigated agriculture in the project area.[35]

An important factor in the farmers' decisions on crops are, in theory, support prices which should function to signal priorities in the choice of crops. In Turkey, until 1973, official prices had been set at harvest time, after the sowing decisions had been made. But even after this had been changed, and the prices have been announced at sowing time, the farmers have sold their crops in advance to the middlemen. Furthermore, during the 1980s, when the official prices were announced half a year ahead of marketing, they fell below the market prices at harvesting time because of the high rate of inflation. The government could not achieve a policy which adjusted the prices automatically in line with inflation, and therefore the support prices had limited influence on the farmers' cropping decisions. Short-term credits have been provided, which should have motivated the farmers to plant other crops, but the procedure for providing these credits has been inadequate, and was found to be unsuitable. Short-term credits are allocated to the farmers at a certain time; they are paid out in one go, and are to be used in the year that they were paid.[36] This rigid scheme has continued until to date, but it is being discussed that it should be changed in the near future. Credits of a certain amount are then to be made available to the farmers who will be able to use the monies at their own discretion. If the credit is not used up in the year of allocation, it can be transferred to the next year. In the year of allocation, however, at least the interest has to be paid back.[37]

Agricultural policies in second crop promotion and diversification also failed until the mid-1980s because yields and net incomes from cotton farming remained high, even in the 1990s, if compared with wheat and maize.[38] Between 1966 and 1971, when the cotton yields/ha in the stage I project area increased from 1,700 kg to 3,800 kg, the support prices developed from TL 1.95 per kilogram, in 1966, to TL 3.6, in 1971, resulting in a sharp increase in the net income per hectare from TL 370 to TL 6,350, which led the farmers to continue to put most of their land under cotton.

---

[35] See T.C. Prime Ministry, State Planning Organization 1989, Vol.2; Yurdakul 1992

[36] The Agricultural Bank is the main creditor; 99.7% of the organized agricultural credits are arranged through this bank. The interest rate is lower and credits have longer terms than on the private credit markets. Credit cooperatives grant credits up to a certain limit to their members only; 95% of their sources are from banks as the cooperatives themselves have hardly any equity capital. Short-term credits for production means have terms of between 9 and 12 months; medium-term credits, with terms of 4 years, are made available for investments in buildings and land. Long-term credits, with terms of 20 years and an interest rate of 45%, are allocated as investment credits, e.g., for agricultural machines. The repayment rate to the cooperatives and to the bank is high (approximately 90%). Since the 1980s, however, the allocation of credits according to purpose has changed: short-term credits have increased, and the medium and long-term credits have decreased. That means that the farmers invest less, due to small profit margins, and live on their former investments. (Bülbül 1995, personal communication)

[37] Bülbül 1995, personal communication

[38] From 1990 to 1993 the net income from cotton yields/ha was fourth times higher than from wheat.

In the years 1971, 1973, 1976, 1979, 1983, 1984, and 1987, high profits could be gained because of high prices, and relatively low production costs (see Table 5.4.). Only in 1973 and 1987 were very high net incomes achieved. In other years, the prices and production costs did not differ substantially, while in 1975 the price per kilogram was lower than the production costs. The cotton production costs ranged between TL 372 and TL 811/ha.[39]

Although maize, soybean and other crops have been promoted with more effective means since 1982, the farmers' interests remained firstly with cotton. This can be explained by the low production costs in comparison with those of high value crops, which, together with the increases in cotton prices, produced net returns from cotton which compared favourably with the net returns from fruit or vegetables. The production costs and the net incomes for selected crops, in 1984, can serve as an example (see Table 5.5.).

The higher production costs of vegetables and citrus, together with inadequate marketing opportunities, provided no incentive to switch to high value crops. Only the farms that are located to the south of the city of Adana planted citrus and vegetable due to favourable marketing conditions (they have easy access to markets and to transport facilities). These farms have an average size of 30 ha, and it is said that the farmers show more initiative, and operate their farms more carefully than the larger holdings. It is believed by local agriculturalists that intensive vegetable production is not suited to holdings larger than a family can handle under Turkish conditions, and it was reported by the Extension Service's foremen that the smaller farmers have been more receptive to diversification because it allows family labour to be spread out throughout the year, and reduces risks.[40]

Since the mid-70s the greatest variable cost item in cotton farming which heavily affected the net incomes, has been the very high share for pesticides. After 1985, it increased to 40% because the input subsidies were reduced under the government of Turgut Özal (see Table 5.6.).

The most remarkable trend has been that due to increasing prices for pesticides, cotton plantations require substantially more monetary inputs before harvesting. Since the mid-1980s the significant increases in the cost of plant protection and decreasing cotton prices, together with the second crop promotion policies for soybean and maize, caused the farmers to give up cotton plantation; maize and soybean have become profitable. The costs for irrigation services, i.e., O&M costs/ha, in cotton plantations decreased from 7% in 1970, to 3.3% in 1989, despite the fact that the water charges were raised after 1984.[41]

---

[39] The production costs/ha for irrigated cotton have been 4.3 times higher than for cotton in dryfarming, but dry-farming is not profitable due to low yields, resulting in higher production costs/kg.

[40] Personal communication with agriculturalists 1995

[41] The O&M water charges were assessed at low levels before 1978, and have approached cost recovery since 1984. The charges for cotton per hectare increased in absolute terms. Only the charges for rice plantation and citrus fruits exceed those for cotton. (see T.C. Bayindirlik ve Iskan Bakanligi; DSI, Genel Müdürlügü 1987, 1988, 1990, 1994, 1995)

Table 5.4. Net income trend in irrigated cotton farming from 1971 to 1988 [42]

Year	Yield [kg/ha]	Production costs [TL/ha]	Index 1971=100	Prices [TL/ha]	Index 1971=100	Net income [10³ TL/ha]	Index 1971=100
1971	3,380	439	100	772	100	1,126	100
1972	3,200	455	104	674	87	702	62
1973	3,400	552	126	1,336	173	2,664	237
1974	3,200	811	185	828	107	51	5
1975	2,200	800	182	757	98	-94	-8
1976	2,900	548	125	945	122	1,150	102
1977	3,100	691	158	773	100	253	22
1978	2,470	597	136	643	83	116	10
1979	3,150	372	85	717	93	1,085	96
1980	2,800	379	86	688	89	867	77
1981	2,560	538	123	637	82	253	22
1982	3,450	435	99	596	77	555	49
1983	2,990	423	96	892	115	1,400	124
1984	3,260	429	98	838	109	1,336	119
1985	2,990	553	126	664	86	332	30
1986	3,080	608	139	700	91	282	25
1987	2,910	514	117	1,171	152	1,911	170
1988	2,810	555	126	785	102	646	57

Table 5.5. Gross value, production costs and net value for selected crops in 1984 (TL/ha)[43]

	Gross value	Production costs	Net value
Cotton	717,500	280,000	537,500
Wheat	212,100	108,000	104,100
Tomatoes	800,000	445,850	354,150
Citrus	995,400	340,000	655,400
Melons	720,000	290,000	430,000

[42] Source: Yurdakul and Ören 1991, p.34. The data used by Yurdakal and Ören are not obtained from empirical research. The revenue figures are average figures for the Province of Adana. A similar problem with the data occurs with production costs and prices. The state purchasing rates are used as a basis, although a considerable number of farmers sold their produce on the free market at lower prices. The fast and non-bureaucratic handling, and payments, to the farmers made the private market attractive as the farmers had to pay back their debts after the harvest. Therefore, net profit figures from 1971 to 1988 only show a trend.

[43] Source: The World Bank 1985, p.46

Table 5.6. Cost items in irrigated cotton farming (percentage)[44]

	1970	1974	1980	1985	1989
Farmland	27	41	20	28	22.2
Fertilizer	14	5	7	8	8.8
Land preparation	3	5	12	3	n.a.[a]
Seed	2	2	2	2	3.5
Plant protection	21	26	35	41	40.8
Irrigation	7	5	6	4	3.3
Harvesting	27	16	17	14	n.a.[a]

[a]   no data available

This item does, however, not include the farm cost for irrigation works, technology, land preparation, and so on, and gives, therefore, an unrealistic figure of the total irrigation costs. Compared with the costs for pesticides, for farmland and harvesting, the cost for irrigation water has remained as a relatively low share of the production costs, and therefore has attracted a lower level of attention in the farmers' decisions, all the more so, the actual repayments attracted little if any attention from the farmers.

The O&M costs incurred by the farmers have, however, been repaid by 10% only. The costs for irrigation services incurred by the farmers in the stage IV project area accounted, until 1993, for only 35% of the normally assessed charges, and since 1993 they have been reduced to 20%. Whether the charges are collected is doubtful because information on whether the lands are irrigated is necessary, and the agency has no control over the actual water withdrawals. The costs for irrigation services changed when O&M responsibilities were transferred to the Water User Associations (see Table 5.7.). The services provided by the state agency, i.e., operation of the main hydraulic infrastructure, maintenance of the main canals, and maintenance of the project's drainage networks, are now completely subsidized while the WUA-members have to fully finance the remaining operation and maintenance expenditures. The WUAs' O&M costs per unit are lower than those incurred by the DSI, but the costs incurred by the farmers will, however, increase because now they have to pay for the services they receive. If they do not, high fines are levied, and non-payers are effectively excluded from receiving services. The costs for irrigation services as a percentage of the production costs in relation to other cost items cannot be calculated as there is no data available. It is assumed that the (calculated) percentage reduces as the WUAs' service costs are lower than those of the DSI.

---

[44] Source: FAO 1988, pp.79-80
In the years 1970 and 1974, insecticides had not been available. The White Fly infestation first occurred in 1974, then again in 1975. In 1976, 10 kilograms of insecticide had been provided free of charge to each farmer which was only sufficient for approximately 5 ha. Therefore, the share of this cost item in 1970 and 1974 is doubtful. (see TU Berlin 1992, p.74)

Table 5.7. Costs for irrigation services incurred by the farmers

Items	First public-farmer setting (DSI/WUG)	Second public-farmer setting (DSI/WUA)
Investment costs		
* delivery and drainage networks	capital charges/ha without interest rate; low rate of repayment	capital charges/ha without interest rate; low rate of repayment
* on-farm drains	subsidized	subsidized
* land levelling	subsidized	subsidized
Maintenance and repair costs		
* delivery networks	only since 1982 the charges have approached the costs; low rate of collection;	WUAs bear the full costs, charged to the irrigators; non-members pay higher fees for receiving services
* drainage networks	charged, low rate of repayment	subsidized
* on-farm drains	nobody is responsible	nobody is responsible
* levelled lands	farmers bear the costs; poorly maintained	farmers bear the costs
Costs for water delivery		
* DSI's services	low rate of collection	subsidized
* WUG's/WUA's services	recovered by beneficiaries; higher rate of payments	recovered by beneficiaries; high rate of payments

3.
Effects of High Groundwater Levels and Salinity on Yields

3.1
Research on Yield Depressions

As was indicated, the cotton farmers enjoyed high net returns, at least until the mid-80s, and even in the 1990s, with rising costs for pesticides, the net returns from cotton remained high when compared with wheat (3.5 times) and maize (2.7 times).[45] Cotton dominated the crop pattern in the whole plain for more than twenty years although several attempts have been undertaken to break its domination. When its decline started in the 1980s researchers from the Çukurova University assumed that the reduction of cotton, which covered more than 80% of the irrigated area until 1980 but only 35% in 1989, and 23% in 1994, was a response of the farmers not only to the increasing costs of pesticides, but to decreased yields caused by high groundwater levels and salinity.[46] From the data available, there is clear connecting evidence between the appearence of pests and cotton yield depressions in 1974/75, 1977/78 and in the beginning of the 1980s. An unequivocal relation between high groundwater levels/salinity and cotton yield depressions is not clearly indicated: the figures used by the researchers are average cotton yield figures which are not only for the project area, but encompass the whole of the Adana province. The cotton yield reductions may also be explained by other factors, of which the continous monocultural cropping is an important one. An early World Bank report underlined that cotton yield depressions coincided with the slackening off of the extension effort, and improper land levelling by the farmers, but also with the increasingly poor maintenance of the drainage canals and the surface drains.[47] Yield depressions of other crops are either caused by climatic conditions or by farm operating practices, e.g., use of fertilizer, irrigation and planting practices. Wheat yields, for example, have been affected by low rainfalls, and citrus yields have decreased due to frosts (see Table 5.3.).[48] After 1982 when the farmers responded to the oilseed promotion policy, some planted maize and soybean as first or second crops. Again, the available data show no clear yield depressions (see Table 5.8.).[49] The average figures for soybean (2,240 kg/ha) and maize (5,180 kg/ha) compare favourably with yields in other countries.[50]

---

[45] DSI, Genel Müdürlügü, Isletme ve Bakim Dairesi Baskanligi 1995b
[46] See Tekinel 1992
One could also presume that cotton dominated because of its tolerance to salt, oxygen deficiencies and groundwater levels, or, in other words, that planting cotton was the farmers' response to high groundwater levels and salinity.
[47] The World Bank 1978, p.17
[48] FAO 1988, pp.88-91
[49] Tekinel 1992, p.7
[50] Adana Tarim Il Müdürlügü Kayitlari (cesitli yillar)

Table 5.8. Second crop yields in the province of Adana (kg/ha)[51]

Year	Soybean Area [ha]	Yield [kg/ha]	Maize Area [ha]	Yield [kg/ha]
1981	7,086	840	-	-
1982	7,400	2,020	3,410	5,010
1983	12,490	2,480	1,755	5,690
1984	18,050	2,890	2,722	7,680
1985	46,516	1,870	18,820	6,570
1986	69,600	2,440	31,800	6,860
1987	59,630	1,950	10,187	4,000
1988	41,500	2,000	16,800	5,000

A study was undertaken by the local scholars, Kanber and Yazar, in drainage-type lysimeters to determine the effects of saline water table-depths on yields of wheat, barley, soybean and groundnut, grown in rotation systems such as 'wheat-groundnut-soybean' and 'barley-groundnut-soybean'.[52] In the experiment, two different water tables with varying salt contents were used throughout the growing season. The authors state that according to the results from the experiment, different water table depths have no statistically significant effects on yields. It was noted that although a water table depth of 75cm decreased the yield of wheat by 19%, the yields of barley, groundnut and soybean actually increased by 8, 19 and 10% respectively. The crop preceeding soybean and groundnut also had an influence on their yields. With a water table depth of 45cm, yields of groundnut and soybean, which had been grown after barley, were respectively 38 to 41, and 72% more than those following wheat. In general, increases in yields were found to be greater with higher water tables, although groundnut yields responded negatively to highly saline groundwater which, at a depth of 45cm, dried out the plants.

The real extent to which yields in the project area (stages I, II and III) are affected by high groundwater levels and salinity cannot, however, be determined from the data and research available, and it is even less possible to evaluate their effects on net incomes. There are, of course, some clues which lend evidence towards yield losses being caused by high groundwater levels and salinity. For example, alongside broken and leaking irrigation canals, the soils become water-logged and crop damages occur. In cotton farming, unskilled and temporarily labourers are employed who receive their salaries on an area basis. They want to irrigate in the shortest time, and they flood the fields which have, in addition, no adequately maintained furrows. Land levelling is not correctly maintained, and ploughing practices have formed local depressions which adversely affect yields.[53] In 1988, in the water subdistricts of Dogankent and Yenice, 83 ha and 197 ha respectively, could not be sown due to saline soils, together with 46 and 200 ha

[51] Source: Adana Tarim Il Müdürlügü Kayitlari (cesitli yillar)
[52] Kanber and Yazar 1992
[53] Personal communication Yazar 1995

which had not been irrigated because groundwater levels were already high. This added up, however, to less than 0.3% of the total service area.[54]

It is assumed here that, within the service area, the negative impacts on cotton yields were, however, limited, foremost, because cotton is a crop highly tolerant to salt, and moderately tolerant to oxygen deficiency and high groundwater levels, and, secondly, because the areas with highly saline conditions were drastically reduced. International research has revealed that cotton yield reductions of between 10 and 20%, which are classified as 'insignificant' by Kovda et al., occur with salinity values greater than EC_e 10,000 micromhos/cm.[55] The areas with saline soils greater than EC_e 10,000 micromhos/cm have shown positive improvements with the installation of the subprojects; the salt affected areas of between EC_e greater than 8,000, and greater than 12,000 micromhos/cm, declined from 85,657 ha (prior to the project) to 25,000 ha (in 1982/84). These highly affected areas are mostly concentrated in the designated stage IV project area. In the upper plain, the groundwater tables are usually lowered by the drainage systems when cotton sowing starts in April; only exceptionally heavy and long-lasting spring rainfall delays the sowing because the water cannot be drained away earlier, due to the insufficiently upheld drainage systems' capacity. The groundwater levels rise with irrigation, and in the summer months, when the salt properties in the soil increase, the cotton plant, which is anyway highly tolerant to salt, has passed the susceptible growth phase. More sensitive crops are wheat, maize, soybeans, and fruits (see Box 5.1.).[56]

Box 5.1. Salt tolerance of selected crops[57]

Cotton is highly tolerant to salt, but shows a moderate tolerance to oxygen deficiency and high groundwater levels.
Maize is sensitive to oxygen deficiency and to high groundwater levels, and is less resistant to salt.
Soybean is moderately tolerant to oxygen deficiency, and is sensitive to salt. With flooding, oxygen deficiency may be the direct cause of damage whereas with high water tables the decreased availability of nitrogen may dominate.
Citrus is moderately tolerant to waterlogging, and is sensitive to salt.

---

[54] DSI, 6.Bölge Müdürlügü, ASO Dogankent ve Yenice Isletme ve Bakim Basmühendislikleri 1988

[55] 10% yield reductions are classified by Kovda et al. (1973, from Achtnich 1980, p.216) as 'insignificant'; 10 to 20% are regarded as being 'low', 20 to 50 are 'moderate', 50 to 80 are 'high', and more than 80% are 'very high'.

[56] See FAO/ UNESCO 1973, pp.280-282

[57] Sources: FAO/UNESCO 1973, pp.265-267, 280-282; Boyko 1966, pp.97-98
The group of highly salt-tolerant crops can stand a salinity of EC_e of about 16,000 to 10,000 micromhos/cm, the group of moderately tolerant crops an EC_e of about 10,000 to 4,000 micromhos/cm, whereas the sensitive group has EC_e 4,000 to 2,000 micromhos/cm.

Soybean reacts with yield reductions of between 10 and 20% at EC_e values of about 6,000 micromhos/cm; vegetables already with EC_e values lower than 2,000 micromhos/cm, and wheat at EC_e values of about 7,000 micromhos/cm.[58] Citrus fruits belong to the most salt sensitive crops, and of these lemons are the most sensitive while oranges occupy an intermediate position.[59]

3.2
Groundwater and Soil Conditions as a Constraint on Double Cropping and on Diversified Crop Patterns

Local researchers have perceived high groundwater levels and salinity as the main constraint on double cropping when the second crop promotion policy for maize and soybean showed success after 1982.[60] Before second cropping commenced, high groundwater levels throughout the year had only been a problem in some parts of the area, and then only in the years when heavy winter rains continued until April. In such years, the farmers have been confronted with inadequate drainage conditions. When second cropping is undertaken after the first irrigation season finishes at the end of summer, the groundwater levels and their salinity content are high. Crops which are moderately tolerant, or extremely sensitive to salt, and to high groundwater levels, cannot then be planted, or are negatively affected. But second cropping is not completely absent in all parts of the plain. The farmers' reluctance to second cropping, and to a more diversified cropping pattern, can be explained by other factors which have been more important in the farmers' decision-making. Major obstacles were undeveloped marketing systems, the lack of processing capabilities, inconvenient credit terms, unstable prices, etc. Until the mid-80s, the main constraint on the further expansion of citrus plantations, for example, had been the disorganized marketing conditions, and not soil and drainage problems. Citrus and vegetable plantations are, meanwhile, widespread near the city of Adana where the natural, and the market conditions are more favourable; in 1994, 13% of the irrigated area was covered with citrus. Even after the abatement of these constraints, second cropping has only slowly expanded as not all farmer have shown interest in double cropping because farm conditions such as labour availability, machinery, and financial resources are unsuitable. Many farmers in the service area have rented out their fields, either because the potential incomes have not been sufficient for their families, or because they have been employed in other sectors. The family-operated middle-sized farms, which are favourably located, have undertaken second cropping with a more diversified cropping pattern because they could spread out their labour forces throughout the year, and thus minimize risk.

Table 5.9. provides an overview of the impact of agricultural policies on farm economy, and the causes for yield depressions.

---

[58] See Meiri and Shalhevet 1973 (p.282), Fig. 3 „Salt tolerance of field vegetables and forage crops"
[59] Boyko 1966, pp.97-98
[60] Tekinel et al. 1985, p.467; Köy Hizmetleri Genel Müdürlügü 1988

Table 5.9. The impact of agricultural policies on farm economy, and the causes for yield depressions

Agricultural policy targets	Instruments	Impact	High groundwater levels and salinity and other yield-depressing causes
since 1963 Change from dry-farming to irrigation; crop diversification; second cropping	Provision of irrigation infrastructure; capital charges without interest; 0&M charges assessed at low levels; on-farm works completely subsidized; subsidies on input prices (fertilizer, pesticides); support prices; credits for second crops; extension service promotes citrus, sorghum, vegetables	High yields and net incomes in cotton farming after 1971 wheat is a profitable, risk-minimizing alternative; high yields and net incomes sorghum promotion ('70): unstable market conditions; citrus promotion ('69/70): disorganized marketing conditions, loss of income in the first years; vegetable promotion: disorganized marketing conditions	Groundwater levels and salinity show positive improvements; concentrations, however, in the southern area; cotton yields affected by pest incidences in 1973/74 and 1974/75; wheat yields affected by low rainfall
end-1970s Wealthier farmers taxed	Fivefold rise in 0&M water charges	High yields and net incomes in cotton farming	Pest affects cotton yields (1977/78)

Table 5.9. (continued)

Agricultural policy targets	Instruments	Impact	High groundwater levels and salinity and other yield-depressing causes
since 1982 Crop diversification; second cropping (vegetable fats and oils); preventing interregional competition over cotton (expected labour constraint)	Provision of irrigation infrastructure; capital charges without interest; assessed 0&M charges approach cost recovery; on-farm works completely subsidized; second crop promotion (subsidies on seed, credits, support prices, marketing); subsidies for fuel and pesticides are cut back	The area devoted to cotton declines after '82 but net incomes still high; cost increase for pesticides; maize and soybean as first crops; second cropping starts	(Stage I, II and III) high groundwater levels are a constraint on second cropping and on crop diversification in some areas (Stage IV) low yields and net incomes
end-1980s Export promotion; exploitation of agricultural potentials	Rehabilitation projects start; completion of incomplete projects; capital charges without interest; sharp cut back in maintenance budget; assessed 0&M charges approach cost recovery; on-farm works completely subsidized; second crop promotion (support prices, credits, marketing); no subsidies for pesticides	Favourable marketing conditions and processing capabilities for citrus fruits, and vegetables; net returns from cotton still high	(Stage I, II and III) constraint is lifted with rehabilitation projects; citrus yields affected by frosts; wheat yields affected by low rainfall (Stage IV) groundwater levels and saline/alkaline soils still high

3.3
Benefits Gained from the Irrigation Project, and Costs Deriving from High Groundwater Levels and Salinity

The benefits from the irrigation projects and the costs arising from high groundwater levels are unequally distributed among the farmers in the Lower Seyhan Plain. In the areas of stages I, II and III, high groundwater levels and saline soils have remained a problem in only a few small locations; for some fields alongside broken and leaking canals, and in years when heavy winter and spring rains have ocurred. The general trend has been that the areas with high groundwater levels and salinity have decreased due to the installation of the subprojects' drainage networks. The farms in these areas have enjoyed high benefits since the operation in the subprojects started in 1969, 1975 and 1986 respectively.

Major differences occur, however, between the farms in the upstream and the downstream areas. The farms in the stage IV project area have not, to date, been supplied with irrigation water (i.e., the project has not yet started). With the subsequent installation of the drainage networks, the soil conditions improved in the first three stages, and the soluble salts have moved and are now concentrated towards the southern part of the delta. The farmers in the stage IV project area have been confronted with the fact that solving the problem in the areas of the first three stages has led to increased problems in the downstream part. In 1984, the stage IV project area (about 40,000 ha) had 2,870 ha of saline soils, 24,455 ha are saline-alkaline, and 4,675 ha are alkaline. Approximately 29,130 ha require leaching and soil improvement measures. 2,676 private landowners have farms of varying sizes in this most affected part of the delta, the Yüregir Plain, where natural topographic conditions make the discharge of excess water, and leaching, impossible until the pumping drainage systems have been installed.[61] The crop pattern in the southern area is dominated by rainfed cotton (43.3%) and wheat (21.4%). (see Table 5.10.) Some farmers have, however, practiced irrigated agriculture, for example with water melons on 6.2%, by withdrawing water from the main delivery canals, the River Seyhan, and from the main drainage canals. In areas with high and also saline groundwater, rice plantation is common (1.8%). About 10,920 ha (or, 27.3%) cannot, at present, be farmed at all due to groundwater and soil conditions.[62] The most important target of the stage IV project is the rehabilitation of these 10,920 ha. The average yields per hectare in the southern area are very low when compared with the yields gained in other parts of the plain: 700 kg/ha cotton instead of 3,000 kg/ha, and 2,500 kg/ha wheat instead of 4,000 kg/ha. (see Table 5.11.) The average annual net income per hectare in irrigated cotton farming in the upstream areas ranges at TL 4.3 million (prices 1990). The average net income per hectare

---

[61] In this area, many small farms of between 1 and 5 ha (i.e., 7.6% of the landowning farm families,) rent out their land, and many larger holdings operate under share-cropping arrangements. 30% of the farm families own farms of between 5 and 20 ha; 28.7% have an average size of between 20 and 50 ha, and about 33.7% own farms larger than 50 ha.

[62] The spatial and timely development of the affected areas cannot, however, be estimated on an accurate data basis.

in the downstream part of the plain is, at TL 79,130, extremely low, i.e., 1.8 % of that upstream.[63] It is estimated that the net income can increase to approximately 1.5 million TL/ha at full development (6 years after completion).[64]

Table 5.10. Present, and estimated crop pattern in the stage IV project area (percentage area)[65]

Estimated		Present	
Cotton	55	Dry cotton	43.3
Wheat	25	Wheat	21.4
Water melons	7	Melons	6.2
Corn	5	Paddy	1.8
Vegetables	3	Unsuitable for cropping	27.3
Sunflowers	3		
Groundnuts	2		
Soybean (second crop)	3		
Corn (second crop)	4		
Groundnuts (second crop)	2		
Sesame (second crop)	1		

Table 5.11. Average yields (kg/ha) and net incomes in the upstream and downstream areas (TL/ha)

	Upstream area	Downstream area (without project)	Downstream area (with project)
Average yield [kg/ha]			
Cotton	3,000	700	3,250
Wheat	4,000	2,500	4,000
Melons	n.a.[a]	12,000	15,000
Maize	9,160	-	5,000
Average net income [TL/ha/a; prices 1990]			
Cotton	4,300,000	79,130	1,500,000
Wheat	1,400,000		
Maize	1,600,000		

[a]   no data available

---

[63] DSI, 6.Bölge Müdürlügü 1990 (p.VII-1), 1984
[64] Ibid.
[65] Source: DSI, 6.Bölge Müdürlügü 1984 (p.VII-3), 1990 (p.VII-5)

The farmers with fields in areas which cannot be cultivated at all could not even achieve the low average net income of TL 79,130 ha/a during the last decades.[66]

4.

The Farmers' Options Towards Groundwater and Salinity Control in Irrigated Cotton Farming

It is evident that the main factors for controlling groundwater and salinity, i.e., investments, operation and maintenance tasks, have been beyond the individual farmers' scope of decision-making. For the downstream farmers, the main constraint to irrigated agriculture is the installation of the drainage networks, including the required drainage pumping stations, as a prerequisite to being regularly supplied with irrigation water. For the upstream farmers where high groundwater levels and salinity have been a constraint on double cropping and on crop diversification, the major obstacle for improvements has been the improper maintenance of the drainage networks, not only at on-project level, but also at on-farm level over which they decide themselves.

It is assumed that if the farmers can increase their net return, they could regard the reduction of water inputs as an individual strategy for the local control of high groundwater levels, waterlogging and salinity. Water inputs could be reduced either by proper irrigation techniques with traditional low-cost irrigation technologies (furrows), or by modern water-saving technologies (drip or sprinkler).[67] Then, positive but limited effects on groundwater levels would emerge, but only if all, or the majority of the farmers belonging to one sub-drainage basin would decide to reduce water inputs by whichever method. If only a few farmers save water, the impact on groundwater levels is neglible. But, as was previously analyzed, reduced water inputs either by changing the water allocation rules, or by means of economic incentives, was not considered by the state agency. The stricter monitoring and enforcement of the water supply scheduling could have prevented overirrigation, and could have positive impacts on the low on-farm application efficiencies.

In the project area, cotton is irrigated by means of furrows with low application efficiencies caused by poorly levelled land, poorly maintained furrows, planting practices, and improper irrigation techniques (e.g., among other things, excess water inputs). Irrigation is carried out by temporarily employed farm labourers who flood the fields in order to shorten their working hours, or to compensate for poorly levelled fields and poorly maintained furrows by applying more water than assigned. The researchers who assumed that water reductions of 25 to 30% would

---

[66] A co-owner of a large holding explains that his family owns some of these unproductive fields. Other fields are rented out, or operate under share-cropping arrangements. His family has shown great interest in improvements, and has started to evaluate commercialized agriculture. The family members belonging to his generation have studied and are engineers employed in industry or commerce.

[67] It is, however, generally assumed that the farmers choose the technology yielding the highest profit, and that they choose the optimal amount of water for each technology.

have no 'significant' impact on yields, consider proper irrigation and planting practices, properly levelled land and maintained furrows, etc. If an individual farmer would decide to apply less water per hectare and crop, then higher costs would occur for land levelling, maintaining the furrows, monitoring the farm labourers to improve irrigation practices, or incentives for the farm labourers to maintain the yields previously obtained. The higher costs are not off-set by higher net incomes because there is no increase in yields, and because, under the present conditions, the area-crop based costs for irrigation water would not decline.

Investments in modern water-saving technologies have not been an option in cotton farming: furrow irrigation is an adequate irrigation technique which also has the advantage that it can be easily maintained. The on-farm efficiency of furrows can be as high as with, for example, basin irrigation or border strip irrigation. Drip irrigation is no alternative in cotton irrigation due to high investment and energy costs. In addition, with drip irrigation, the leaching of salt properties in the soil is not guaranteed. The use of sprinkler irrigation systems faces high investment and energy costs. Under the climatic conditions in the plain it is presumed that sprinkler irrigation favours higher incidences of pest. To avoid losses through evaporation and adverse effects on pollination during full blossom, night irrigation is recommended which would increase the irrigation costs for the farmers (costs for labour). The adoption of input-conserving irrigation technologies in cotton farming, be it drip or sprinkler, are, however, not correlated with higher yields and net returns. Investments in water-saving irrigation technologies would have caused a shift in the cost structure, leading to a reduction of the net incomes due to non-increasing yields, all the more so the costs for water would not reduce. According to Achtnich, irrigation using modified furrows with shortened runs can be regarded as an adequate alternative in cotton farming: the fixed costs per hectare increase little; a higher yield is attained, together with an 8 to 10-day shorter growing period, and a water saving of 25 to 30%.[68] This alternative would require a centralized approach with regard to the water distribution method, and the assessment of the water charges.[69]

It was previously analyzed that the farmers had shown no interest in maintaining the state-owned and state-constructed on-farm drainage systems, and in investments in additionally required on-farm drainage systems. According to the recently discussed investment approach for on-farm development works, the farmers, who become the owners of the on-farm installations, are responsible for their maintenance. There are, however, well founded doubts that all farmers are able, or willing, to invest: some farms are not large enough to obtain a sufficient net income for all family members; some families are no longer interested in agriculture because their members are employed in other sectors; other farmers lease fields or rent them out; under the various share-cropping arrangements, neither the tenants nor the owners have shown an interest in investments. If investments and maintenance depend on the individual farmers' decisions, they may also decide to save

---

[68] Achtnich 1980, p.492
[69] Volumetric pricing is, however, not a necessary condition. Water charges could be assessed in relation to, and varying with, the applied technology, the time duration of relase or other proxies.

money through non-investment and non-maintenance, because the additional costs may exceed the anticipated benefits which derive from them. Because the negative impacts of non-maintenance are not only limited to those farm units directly concerned, i.e., the negative consequences will also affect others, solutions to the maintenance problem require incentives which induce collective action. It seems to be necessary to limit the individual farmer's range of decision-making: the farmers would sign long-term contracts with the state agency which clearly indicate their responsibilities, liability for damages, the maintenance instructions to be followed, the time intervals, the monitoring, etc. This option has the advantage that if the farmers are not able, or willing, to finance the investments, they are nevertheless responsible for their maintenance, with the agency exercising control over the proper and regular implementation of maintenance works. The farmers would have to commit themselves to these responsibilities before construction commenced.

5.
Conclusions

More than 20,000 private farm units of varying size and structure are located within the project area. The majority benefited from public investments and subsidies, and as they are also negatively affected by high groundwater levels and salinization, it was assumed that they would show an active interest in control measures.

1. The benefits from the irrigation project, and the costs deriving from high groundwater levels, were unequally distributed among the farmers in the project area. Negatively affected farmers have no control means.

The analysis has shown that the cotton farmers in the stage I, II and III project areas have enjoyed high yields and net incomes for almost 30 years, and that high groundwater levels and salinity had little negative impacts on farm economy within the service area. In some years, cotton yield depressions occurred due to high pest incidences and monocultural cropping; the introduction of higher value crops and second cropping was, above all, hampered by marketing conditions and processing capabilities, and not by groundwater and soil conditions.

Major differences, with regard to the costs deriving from groundwater levels and saline/alkaline soils, occurred between the farms located in the upstream and downstream areas. In the areas of stage I, II and III, high groundwater levels and salinity have only been a constraint on double cropping and crop diversification for some farmers because the drainage systems were not properly maintained. In the southern areas approximately 11,000 ha (i.e., 6.3% of the total project area) cannot be farmed at all due to groundwater and soil conditions. There, the main obstacle on irrigated agriculture has been the lack of investments in the main drainage infrastructures, including the required drainage pumping stations. It can be concluded, however, that the demand for controlling groundwater and salinization has been weak because the farmers are not all, and in the same way, affected. The farmers who are most severely affected could only exercise influence on political decisions either through the mechanisms of the political system, or through their interest groups. However, the implementation of the major control measures has been beyond the farmers' scope of decision-making.

2. Incentives to induce collective action for the control of high groundwater levels and salinization at on-farm level have remained poor.

The case study has shown that incentives, be they economic or administrative, to induce collective action have remained insufficient. No incentives have been set to instigate the rationing of water: water charges are area-crop based, and provide no incentive for the farmers to save water either by proper land levelling or irrigation practices. Water inputs could be reduced by the main system's operation activities, which was not considered. However, to obtain the yields, proper irrigation techniques and land levelling would have imposed higher costs on farm economy. Water-saving technologies have been no option in cotton farming, and irrigation by means of modified furrows with shortened runs requires a central approach to fit in with the water delivery method. Taxes on drainage effluents, be it taxes on outputs or on inputs, have not been considered, and cannot, however, be applied. While the ouput-oriented approach faces great difficulties (e.g., problems of monitoring and measurement), prerequisites to applying the input-oriented approach do not exist (i.e., to assess the taxes per unit of water used, the volume needs to be measured). The major incentive for water saving in cotton farming would have to emanate from changes in the water allocation rules and operation activities; other approaches would not have been effective.

Central groundwater and salinity control measures, e.g., investments in drainage systems and their proper maintenance, have not been implemented due to insufficiently provided funds, and both public-farmer settings provide no satisfactory solutions to the drainage systems' maintenance, neither at on-project nor at on-farm level. The area-crop-based water charges which are, however, the costs for irrigation services, can have effects on the farmers' cropping decisions but not on water conservation, or water saving, because it leaves the marginal cost of water to the farmer at zero. The only means to put water savings into effect, and to instigate the farmers to reduce water inputs, will be a system of water prices.

Chapter Six
Discussion and Outlook

The analysis of the Lower Seyhan Irrigation Project has aimed to reveal why effective means for controlling high groundwater levels, waterlogging and salinization have been ignored by the participants, and why they have participated in activities which have produced high groundwater levels and salinization. The basic assumption leading the analysis has been that the incentives facing the self-interested individuals, be they politicial, bureaucratic, civil or economic agents, have biased towards the control of high groundwater levels and salinization. The task of the case study was to prove whether the Institutional Rational Choice approach can be applied for the analysis of the environmental problem in a large-scale public irrigation scheme. This approach makes the general assumption that institutions, or rules-in-use, decisively influence the individuals' decisions and the respective outcomes; the participants would behave rationally within a given institutional setting, and the rules would provide (dys)functional incentives for the participants involved.

Irrigation systems are common-pool resource systems, i.e., they share characteristics with private and with public goods, where the delivery infrastructure is jointly used by the beneficiaries, and where the substraction of resource units as input factors causes externalities for other users. While the interdependence created by the joint use of the irrigation system is studied intensively, the institutional analysis of problems associated with a project's drainage networks, i.e., 'collective goods', which are jointly used and where exclusion is costly, has remained poor. The same holds true for research on on-farm drainage systems, which are primarily subject to private use, and which need to be installed and maintained by a majority of the farmers in one sub-drainage basin to secure effective results for controlling groundwater levels. However, both interdependencies create the potential for inefficient resource use if they are not mediated through enforceable rules which allow not only for the provision of delivery and drainage infrastructures together with their maintenance, but also for controlling water allocation, especially if free-riding behaviour is not effectively eliminated.

The Lower Seyhan Irrigation Project is characterized by a large service area and a great number of private farm units; financially dependent state bureaucracies were responsible for the planning, design and construction of the project. Two types of water user organizations have contributed towards operation and maintenance activities in which they have shared responsibility with the operating state agency to supply services for numerous private farmers who use irrigation water as an input factor in agricultural production. The management unit of a large-scale irrigation scheme faces specific problems which derive from the size, e.g., information processing and transforming O&M activities, difficulties of internal control, and rule enforcement. The presumption of this study is that these problems would be minimized by introducing farmers' organizations at the lower level of the irrigation schemes to participate in O&M tasks. Their participation would improve water allocation along the lines of similations with small-scale irrigated common-property regimes, i.e., farmer-owned and -managed systems. This assumption departs from empirical research on common-property regimes carried out by scholars of the Institutional Rational Choice approach who regard irrigated common-property regimes as alternatives for state-managed common-pool resources. Their research has revealed that small-scale irrigated common-property regimes show advantages over large-scale public schemes: water supply becomes more predictable; efficient water allocation and water-use can be guaranteed under changing environmental circumstances; the benefits to the individual are equal to the cost; the officials are accountable to the beneficiaries, and free-riding is limited by monitoring and strict rule enforcement. The main reason for these advantages is that they provide proper incentives, when compared with publicly managed schemes. There, a multiplicity of decision-makers with differing and conflicting preferences is involved. In large-scale public irrigation systems with a highly centralized institutional setting, those deciding on the provision are not those who manage the system, or who use it; the managers are not accountable to the beneficiaries; their financial contributions towards the production costs have almost no influence on the budget nor on the performance. A problem of major concern is how the state and the management unit can ascertain collective action among interdependent users by defining adequate and enforceable rules for the allocation of water and for the control of high groundwater levels and salinization. This is particularly important in the case of infrastructure-intensive resources; in the case where many transformation and transaction activities are necessary to achieve an adequate, reliable and equitable water supply and to sustain the irrigation and drainage infrastructure, and in cases with a large number of polluters and victims. The problem of inducing collective action is essentially one of establishing and enforcing institutional rules. The long chains of communication and decision-making in central institutional arrangements with dependent rule-making and enforcing authorities cause losses of information, which lead to difficulties in control. The costs resulting from a lack of information, or from the opportunistic behaviour of the participants add to the already high cost of information processing. In jointly managed large-scale irrigation systems which incorporate participating user groups to varying degrees, there may be, depending on the model, more than one rule-making and rule-enforcing authority, and information processing and control may become more effective. A central point is whether jointly managed large-scale irrigation systems have the potential for implementing effective groundwater and salinization control means, because demand for controlling salinization can effectively be addressed.

In the following section it is discussed whether Elinor Ostrom's concept could be successfully applied for analyzing the given setting and the relevant object, and the empirical results obtained from the case study are briefly summarized (1.); the factors causing high groundwater levels and salinization are grouped, and problems of cooperation and coordination are differentiated (2.); the conditions for changing the institutional arrangements and for the successful participation of user groups, another theme of the Institutional Rational Choice approach, are revealed (3.). Finally, departing from the experience of the Lower Seyhan Irrigation Project, working hypotheses are formulated towards an effective strategy for controlling groundwater levels and salinization in jointly managed large-scale irrigation systems (4.).

1.
The Concept for Analysis and the Empirical Results

For analytical purposes, Elinor Ostrom's concept of action arenas has proved to be useful to address the multiple levels of decision-making with their changing participants, and their varying sets of rule-ordered relationships. With reference to Ostrom[1] and Ostrom et al.[2] several action arenas have been defined: the <u>first</u> is the planning and design process where two state agencies cooperate; in the <u>second</u> field, investment decisions are made, including regulations on the beneficiaries' financial contributions. Production decisions, i.e., the act of executing investment and generating O&M services, are taken in a <u>third</u> and <u>fourth</u> action arena; in the fourth field, transformation and transaction activities are relevant, and the beneficiaries' joint use of the project's irrigation infrastructure requires the inducement of collective action. The sustainment of the project's drainage networks is part of this arena which requires particular attention. Decisions in farm economy constitute a <u>fifth</u> action arena. The modelling of this arena deviates from Ostrom's proposal for analyzing infrastructure development, and its sustainable use, but it is consistent with her method. In this field, water resource units serve as a central input factor to which others have to be adjusted to gain maximum advantage, and agricultural effluents are produced by private farm-firms, from which they are negatively affected. Investment decisions in the on-farm drainage system, in land levelling, in water conservation measures, and in the maintenance of on-farm drains are important issues to prevent high groundwater levels and salinization, and the modelling of this arena reflects the fact that these decisions are made by private decision-makers. The inducement of collective action at the farm levels for effective groundwater and salinization control measures needs to be considered as a separate issue.

Elinor Ostrom's set of rules, which configures an action arena, indicates the most important and decisive points for analyzing a multiorganizational arrangement, and the problems of cooperation as well as coordination in multiorganizational arrangements can be adequately analyzed by applying her set of rules, e.g., boundary rules, authority rules, rules indicating the positions, the pay-offs, and aggregation rules, etc. As a result of the case study, the relevance of internal and

---

[1] See Ostrom 1986a

[2] Ostrom et al. (1993, pp.31-34) consider subsequent phases of infrastructure development, i.e., planning and design; finance; construction; operation and maintenance, and use.

external variables should be emphasized. These variables affect the individuals' strategies, the decision-making process, and, finally, the outcomes in different ways. If the outcomes of one action arena create conditions for the decision-making in another arena, they are considered as *external* variables over which the participants to an action arena have no control. They decisively influence the individual strategies, and they exercise influence on the individuals' decision-making, either as economic incentives, regulation standards, legal regulations, etc. The *internal* variables reflect conditions which are inherent to the decision-making process within one action arena. Table 6.1. illustrates the important external and internal variables in the various fields. It gives a clear picture that in all arenas multiple participants with conflicting preferences are involved. Even within one action arena there are varying sets of participants deciding on interconnected items having no direct relations with each other. Decisions made in one phase create *external* variables not only for the following phases of decision-making but also for the participants' strategies in other arenas. The great number of participants, e.g., political units and public agencies, civil organizations such as the water user groups and the water user associations, and private farmers, shows the need for procedures to induce cooperation and coordination with regard to interdependent and interlinked issues. Although decisions and actions from within one action arena affect the participants' strategies of the subsequent phases of decision-making, it is a characteristic of the setting that there are no institutionalized accountability linkages, i.e., the participants have no direct means of control over these actions. (see Table 6.1.)

First arena: Planning and design

Both of the involved planning agencies recognized the importance of controlling groundwater levels and salinization to make full use of the agricultural potentials of the project area. The two state agencies successfully cooperated to apply for investment funds in detailing their work based on legally defined responsibilites. The project proposal, part of which was the technical outlay for an intensive drainage system, was supported by the international donors who emphasized the importance of adequate drainage systems. The design of the delivery network considered an estimated cropping pattern which was not met by the farmers until the middle of the 1980s, and caused the delivery systems' capacity to be insufficient in times of peak water demand.

Second arena: Investment decisions

Adequate investment funds from external and internal resources were provided for the first three project stages. Investments on 40,000 ha in the fourth stage were stopped by political decision, made when the capital costs of the drainage networks became extremely high. Access to external funds was impossible, and internal resources were directed towards other sectors, regions and projects.

Table 6.1. Action arenas in a multiorganizational arrangement

Action arena	Items	Participants	External variables	Internal variables
Planning and Design	Project objectives Technology Management	Planning units (irrigation and agricultural agency) Consultants	Requirements from foreign creditors Targets for sector policy	Number of decision-makers Heterogeneity of preferences Coordination procedures Decision-making authority Skilled personnel
Investments	Budget Assessment and collection of capital charges Executing units	Public decision-makers Legislature Farmers Creditors	Past decisions on sector policies Requirements from foreign creditors Proposed plan	Number of public and private decision-makers Heterogeneity of preferences Coordination and cooperation procedures Decision-making authority
Executing Investments	On-project installations On-farm installations	Public agencies Private construction firms Farmers Creditors	Past decisions on project size and design Past decisions on financing strategies Available funds	Number of public and private decision-makers Heterogeneity of preferences Coordination among public agencies, with the private sector and cooperation with beneficiaries Decision-making authority Skilled personnel, equipment, labour productivity

Table 6.1. (continued)

Operation and maintenance	O&M budget Assessment and collection of water charges	Political decision-makers Public officials	Requirements from foreign creditors Past decisions on financing irrigation services O&M budget proposals	Number of political, public and civil decision-makers Heterogeneity of preferences Coordination and cooperation procedures among public agencies, with WUGs and beneficiaries Decision-making authority Skills
	Water supply planning and operational activities Maintenance programmes and activities Cooperation with WUGs and control	O&M unit of the state agency	Legal regulations on responsibilities Past decisions on water charges and annual budget Requirements from foreign creditors Technology and water availability Crop decisions	
	Contributions to and organization of O&M Water charges collection	WUG with village mayors, ditch-riders and hired personnel	Legal regulations on formation, responsibilities, control and financing	
Use of water at farm level	Water withdrawal Financial contributions Water conservation	Farm operators Farm labourers Water User Groups Irrigation agency Agricultural agency	Past decisions on crops Water charges, input and output prices Water availability System operation	Number of private and public decision-makers Heterogeneity of preferences Coordination and cooperation procedures Decision-making authority Available inputs, skills

The costs for expropriations were kept at a low level by constructing steeply sloping sides in the main drains. This decision had important implications for subsequent maintenance activities, e.g., additional stabilization measures to prevent silt-inwash. The higher investment costs for the concrete-lined deliveries, including canalettes, reduced the percolation of water and, at least for a while, the maintenance costs, but later in time increased, in particular, the costs for repair. Technological innovations and additional investments to correct for the insufficient capacity of the deliveries were not considered. In this field, non-cooperation with the farmers to adjust the technical system to varying local conditions showed negative results. In addition, the continious land fragmentation made direct access to the deliveries impossible for some farmers, and, as further investments for the already dense delivery netwerks were omitted, the enforcement of rights-of-way became an important, and difficult, task to secure reliable water supplies.

Decisions on investments, be they technological innovations within the service area or the completion of the main drainage system in the stage IV project, were outside the control of the irrigation agency and the farmers. The decisions on investment funds and on capital charges are made by different public decision-makers, and two different investment strategies were implemented for the engineering works at the on-project and the on-farm level which had important implications on the execution of investments. The investment funds provided showed no relation to the collected capital charges, and vice versa. Further deficiencies arose because not all beneficiaries were charged to contribute towards the investment costs, e.g., the industries and cities; the assessment procedure caused high subsidies; the charges were poorly collected by the staff of the Ministry of Finance; their collection also encouraged free-riding behaviour.

Third arena: Executing investments

The uncoordinated efforts of the two state agencies caused the levels of the projects and the on-farm drainage systems to be unadjusted, which had to be corrected entailing high costs. The heavy delays in the installation of the on-farm drainage networks was directly related to the investment strategy for on-farm works. Therefore, the foreign creditors initiated a change in the investment strategy which was supported by the irrigation and agricultural agency. Prior to this initiative, the on-farm drainage systems should have been financed by the farmers, either from their own resources or with credits from the agricultural bank. The farmers showed, however, no interest in investments because their net incomes from irrigated cotton farming were already high, and there were no urgent needs to invest in drainage systems; funds for credits were insufficient and only available with inconvenient conditions. The investment strategy which is in force at present, completely subsidizes the installation of the on-farm drainage systems. However, the publicly financed technical systems at the farm levels remained inadequate unless the farmers installed additional subsurface drains in between the state-financed and constructed drains. These additionally required investments were also omitted by the farmers. Irrespective of the investment strategy, cooperation between the constructing units, i.e., the public agency and/or the private firms, and the farmers was poor. Therefore, the engineering works could not be organized in a rational way, causing delays and the overrun of the estimated project costs.

Fourth arena: Operation and maintenance

High water inputs into the project area accrued, apart from unrepaired cracks in the canalettes, from the calculated and released amounts of water per crop and hectare, the central part of the operation activities of the state agency; from the higher than assigned water withdrawals by the farmers, because the Water User Groups could not ensure the enforcement of the water supply schedules; and from irrigation with saline water in the stage IV project area, which was legalized because the state agency could not prevent it.

Soil conservation and water-use at the farm level is the legal responsibility of the agricultural agency, who conducted research on crop yields in all local soil series with the result that about 20% of the applied water could be saved without significant effects on yields. In this field, the competing views of the two agencies were not resolved. The weak instruments of the agricultural agency, i.e., that it can only advise farmers, caused continiously high water inputs. Cooperation did not emerge, and water savings were not considered by the other participants.

Due to their institutional setting, the Water User Groups could not effectively enforce the water supply schedules: the ditch-riders' monitoring capacity was limited as the number of ditch-riders was small, and they were confined to only acting within one village. The imposition of sanctions and the settlement of conflicts by the headmen was impeded by their conflicting role as politically elected mayors, and administrators of the user groups. Legally imposed sanctions were, however, not very harsh. The irrigation agency did not have the power to seriously penalize rule-opposers, and monitoring and enforcement costs were high. Its main tasks were, however, the planning and the daily implementation of the water supply in the main delivery system which was, anyway, inadequate due to the low labour productivity and non-available money for extra salaries. Budget shortages negatively affected the agency's operational activities because the working time of the personnel could not be synchronized with the technical system.

Routine maintenance programmes were set by the irrigation agency, and local maintenance requirements were, at least to some extent, considered insofar that they refer to the manual works done by the Water User Groups. The main problems occurred with the activities which were within the responsibility of the state agency, i.e., the maintenance of the drainage networks and the repair of the deliveries. Some activities, i.e., chemical, biological and mechanical, could not be executed by the Water User Groups due to the lack of skilled man-power and equipment, and the contracts concluded between the state agency and the user groups thus did not include drainage affairs. The maintenance activities in the project drains were most severely affected by budget shortages, and the maintenance of the deliveries, although better than that of the drains, has also been affected since the 1980s, and thus waterlogged fields alongside the channels, caused by cracks and leakages, are widespread. A particular problem has been the maintenance of the state-owned on-farm drains: nobody has been made responsible for their maintenance, which has led to them being completely neglected.

The provision of adequate maintenance funds is outside the control of the irrigation agency and the farmers. The deficiencies occurred because not all beneficiaries were charged to contribute towards the maintenance costs, e.g., the industries and the cities. The ex-post facto assessment procedure caused high subsidies, and the charges were poorly collected by the staff of the Ministry of Finance, encouraging free-riding behaviour. The allocated budgets showed no relation to the collected charges, and vice versa. The assessment procedure for O&M water charges, and the way they were collected, demonstrated a structural weakness which caused a situation in which the recurrent costs could never be met. There were no effective sanctions on non-payers, and the lack of cooperation and coordination between the Ministry of Finance and the irrigation agency created no incentives for the farmers to pay the O&M costs. On the other hand, the Water User Groups were able to collect a high rate of their part of the O&M water charges due to the closer linkages between the Water User Groups and the irrigators, and due to the control they exercised on repayments.

The maintenance of the deliveries and the water supply, including the enforcement of the scheduling, improved with the transfer of O&M tasks to Water User Associations. The main reasons are the increase in labour productivity; heavier sanctions on non-payers and on rule-opposers are imposed and enforced; an adequate budget is constituted by the beneficiaries' payments, and the repayment mode fits in more with the farmers' ability to pay. It cannot yet be judged whether improvements have been made, or will be made with the state agency being responsible for the project's drainage system. These services are completely subsidized and depend on central budget allocations, and the state-provided services are not part of the contracts concluded between the Water User Associations and the state agency.

Fifth arena: Groundwater and salinization control at the farm level

Despite deficiencies in investments, in maintenance and in operation, groundwater and soil conditions improved within the service area with the result that the negative effects, and thus the pressure, on farm economy was limited. High groundwater levels and saline/alkaline soils are concentrated in the stage IV project area, and in some parts of the stage I, II, and III they have been an obstacle to second cropping since the mid-80s. The main difference has occurred between the upstream and downstream farmers: while the upstream farmers have enjoyed high net incomes from irrigated agriculture, the downstream farmers bear the costs from inadequate groundwater control means, and from the postponed investments in the not yet completed delivery networks. But even in the downstream area some farmers have practiced irrigation. Only a small group of farmers whose fields comprise of 10,000 ha, cannot practice agriculture at all due to groundwater and soil conditions until heavy investments in the drainage networks and in soil reclamation have been undertaken. In the analyzed case the main solution to high groundwater levels and salinization has been investments in the drainage networks at on-project and at on-farm level and their maintenance; investments in delivery systems with low percolation coefficients and their maintenance, and only then reduced water inputs.[3]

At the farm level where water is used as an input factor in agricultural production, water savings have not been considered. Policies encouraged irrigation and certain crops by setting low water charges per crop and area, which was not effective. The farmers responded positively when irrigation by means of furrows was recommended because higher yields were obtained with these low-cost innovations. Later in time, the farmers invested in water-saving technologies, i.e., sprinkler and drip, when they planted other crops with a higher profitability. This caused higher investment costs but reduced labour costs. For the farmers, the dominant decision-making factor has been the profitability of the crops; decisions on water have been secondary. The costs for O&M were low if compared with other cost items in irrigated cotton farming. The O&M charges have been area-crop-based which has provided no incentive for water savings, and they were, however, poorly collected. The farmers' non-payments towards the O&M costs had no consequences on the services provided, nor on the O&M budgets allocated, which again was no incentive to induce the farmers to pay. Negative effects of poorly maintained furrows and inadequately levelled land on yields are counter-balanced by the farm operators, or the farm labourers, with higher water inputs; or to put it differently, higher labour costs are substituted with higher water inputs at zero cost.

The farmers omitted investments in the on-farm drains, and the maintenance of the state-owned networks, for which more than one reason can be assumed: the negative effects caused by high groundwater levels and salinization have been limited, and yield reductions within the service area were not too severe. Second cropping was impeded by the poorly maintained and insufficiently installed on-farm drains, but improvements in this field would have made corrections in the main drainage systems necessary to achieve a technically effective result, which was outside the farmers' control. The costs for installing the drains and for their maintenance may not always be off-set by increases in the net incomes. Under the share-cropping and leasing systems, investments depend not on the cultivators but on the owners. Maintaining the farm drains under share-cropping and leasing systems, with contracts only lasting for one or two years, would have caused costs which might not have benefited the share-croppers or leasers who maintained them. And, finally, the installed on-farm systems are state property in which the farmers might not want to invest.

2.
Coordination and Cooperation in Multiorganizational Arrangements

From the empirical analysis, the deficiencies which caused the generation of high groundwater levels and the salinization of soils can be grouped: the *first* group

---

[3] For details, see the conclusions at the end of the Chapters Three and Four

comprises of policy decisions in a narrower sense which decisively determined that the problem was not adequately addressed, and which, on the contrary, set conditions and incentives for both groups, the state agency and the farmers, which induce the emergence of high groundwater levels and salinization. The issues were decisions on investments, on maintenance funds, on water rationing, on the assessment of O&M water charges, and on the maintenance of the on-farm drainage systems. The *second* group of deficiencies stems from the structure of the multiorganizational arrangement. This arrangement can be viewed as a 'policy output' where the behaviour of the individuals failed to have the effects policy-makers, or bureaucrats, wanted to achieve, as was the case with investments in the on-farm drainage systems, the recovery of recurrent costs, and the Water User Groups' contributions towards O&M tasks. The failures to achieve the intended results are rooted in the lack of an adequate incentive system. A *third* group of deficiencies, also non-intended, derives from the large size of the irrigation project where, due to the high costs of monitoring and enforcement, the state agency could not secure collective action although it was supported by Water User Groups.

The empirical analysis indicates that in a multiorganizational arrangement the success of implementation, operation and maintenance, for reliable water supply, for controlling groundwater and salinization, etc., depends on cooperation as well as on coordination, which are both part of the collective action problem. Cooperation is demanded when there are interdependent relationships among the individuals, and when the action of one individual, or one organizational unit, is also contingent on the actions of other individuals, or units. This has been the case in the process of installing the on-farm drainage systems, where public or private enterprises need to cooperate with numerous private farm-firms; in the case when the poorly, or non-maintained on-farm drainage system affects other farm-firms. Coordination becomes necessary if two or more administrative units act on behalf of a public policy, as was the case with the construction works at on-project and on-farm level implemented by two public agencies; or the activities carried out by the Ministry of Finance which were uncoordinated with the irrigation agency. Both cooperation and coordination among organizational units do not arise spontaneously, equally so cooperation amongst individuals. Coordination between public administrations needs to be addressed by political decisions and requires an appropriate mode, or rules. From the case study it is indicated that coordination across the different organizational units is not spontaneously forthcoming, and is more difficult to achieve than within one administrative organization. The competing views of the irrigation and the agricultural agency with regard to adequate water releases can serve as an example. Coordination would have saved water and reduced the high water inputs. Although the responsibilities have been legally defined, and power has been assigned to the agricultural agency, the irrigation agency has commanded the on-farm water use. Its activities have been more influential, and the means available to the agricultural agency have been weak. Coordination would have been costly for the irrigation agency, making the process of water supply planning and its daily implementation increasingly complicated. On the other hand, the information available to the agricultural agency has remained unexploited. From the case study it can be assumed that the main forces which impede coordination across the administrative units are their attitude to secure a domain of power; highly centralized structures of decision-making impede coordination between the lower administrative levels; the processes of bargaining and co-ordinating impose high costs on the participants, and, finally, is the simple but important fact that procedures for coordination are not institutionalized, for whatever reason.

3.
The Conditions for Changing Institutional Arrangements and for Successful Farmer Participation

The empirical analysis has shown that major changes in the institutional arrangements have been undertaken. The empirical process has verified the assumptions of the Institutional Rational Choice approach that if transaction costs are prohibitively high, and if free-riders cannot be excluded, a change in the institutional arrangement is desired to provide an appropriate incentive system.

The introduction of Water User Groups positively affected the operation and maintenance of the irrigation system. On an empirical basis, the main factor which has promoted, and prompted, this innovation has been the recognition that the state bureaucracy was not able to supply irrigation services, as requested by the farmers, without user groups if the service area is large and if the number of irrigators is great. In addition, their establishment can be interpreted as the wish to limit the size of bureaucracy. The Water User Groups were initiated by the public irrigation agency; these groups were based on the village administration units who could secure strong leadership as they were elected by the majority of the farmers. Although this concept can be regarded as an extension of the existing public administration structure, it provided incentives, or advantages, for all parties involved, i.e., the irrigators, the irrigation agency, the national treasury, and the village administration units themselves. The relationship between the state agency and the Water User Groups were clearly defined, and contracts were concluded over their responsibilities, their range of discretion, and their cooperation with the state agency. However, the incentives provided for the village administration units caused them to make suboptimal contributions towards operation and maintenance, and the contracts did not determine drainage affairs. The Water User Groups were accountable to the state agency, and not to the irrigators, and, therefore, the irrigators' demand for the control of salinization could not directly be expressed.

Later in time, O&M responsibilities were transferred to Water User Associations, improving the operational efficiency and the maintenance of the delivery systems, with the result that the water supplies have become more predictable and that the irrigators' user-rights to water can be guaranteed to a higher level. The initiation of the transfer programme expresses the fact that the state has not been able to limit free-riding behaviour, which has caused irrigation services to be highly subsidized under increasing fiscal constraints. The transfer of O&M responsibilities to Water User Associations was promoted by an influential external participant, i.e., the World Bank, but it was supported by most of the participants involved. It solved, at the same time, disagreements between the Turkish government and the World Bank on the recurrent cost recovery issue, and, after this disagreement was settled, the Turkish government could achieve support for the rehabilitation project and the stage IV project, which both started in 1989. After the transfer of O&M responsibilities to Water User Associations, the total budget allocations and indirect subsidies could be reduced, although the subsidies for the maintenance of the drainage remain. The DSI's establishment was not negatively affected, e.g., its personnel was not laid off. As the owner of the infrastructure it supervises the WUAs, and retains the authority to cancel an association. Its remaining responsibilities are the maintenance of the drainage systems, and the operation of the main headworks and the main channels, and it provides extension services for the WUAs on O&M tasks. Although the O&M costs incurred by the farmers are de-facto higher than under the DSI's governance, e.g., now the irrigators have to pay the charges and free-riders are effectively excluded, the irrigators benefit from the higher performance standard due to the increase in labour productivity. The enforcement of the water supply scheduling is more effective, and sanctions on rule opposers are executed. The major incentives for the irrigators to join the WUAs as members are that the services are more expensive for non-members, and non-members receive water last in line.

The main factors which decisively influence the transformation of institutions are shown in Table 6.2. In general, it can be said that external institutional inputs were only effective when they were perceived as viable solutions by the relevant and decisive participants. The internal variables are not of equal importance. The irrigators' perception of the current conditions, and their perception of the expected benefits and costs deriving from the proposed rules in comparison with those prevailing, and the existence of initiative public and private leaders are of major importance.

Table 6.2. External and internal variables affecting institutional change[4]

External variables are
• conditions set by foreign creditors

Internal variables are
• number of decision-makers
• heterogeneity of preferences
• rules-in-use for transformation
• initiative leaders
• current conditions including costs from prevailing rules
• proposed rules
• expected benefits and costs as perceived by the participants if compared with the status quo

[4] Modified from Ostrom 1990, p.199

The analyzed case has shown that although the farmers objected to some forms of participation at a certain time in the past, they were willing to participate at another point in time. The incentives for participation must respond, in time and circumstances, to specific conditions (e.g., their ability and willingness, their perception of benefits and costs, their experiences with cooperation, and the building of mutual trust). In this respect time dimensions have been important for successful user-participation as they essentially affect the individuals' perception of benefits and costs if compared with the status quo. (see Table 6.3.)

The establishment of Water User Groups and of a joint management structure with Water User Associations have been successful as they adequately, and differently responded to the irrigators' willingness to participate. Their willingness for participation essentially depends on immediately visible benefits, i.e., adequate and reliable water supplies. User groups have only recently contributed, at least partly, towards groundwater and salinization control. But these contributions are still minimal, and they are better characterized as a by-product of the activities which guarantee adequate and reliable water supplies, i.e., the proper maintenance and repair of the delivery system.

Table 6.3. Phases of participation

First phase (1971-1981): Irrigation Cooperatives and Irrigation Associations
initiated by the external donor;
supported by the government;
supported by the state agency;
not supported by the majority of the farmers.
Outcome: Only a few were established.

Second phase (1981-1994): Water User Groups
initiated by the state agency;
supported by the government;
supported by the village administration units;
supported by the majority of the farmers.
Outcome: Water User Groups were established in almost all villages.

Third phase (1994 continuing): Water User Associations
initiated by the external donor and the state agency;
supported by the government, the majority of the irrigators, the village administration units;
not supported by a minority of farmers.
Outcome: Water User Associations have been established over a very short time, and cover the entire service area.

4.
Towards Coordinated Management of High Groundwater Levels and Salinization in Large-Scale Irrigation Systems

Recent discussions focus on the negative impacts of state bureaucracies managing irrigation systems, and the predominant trend relies on the transfer of large-scale public irrigation systems to user organizations. This emphasis stems from the positive experiences with small-scale farmer-owned and -managed irrigation schemes with management structures providing adequate incentives to improve the operational efficiency. But these transfer programmes do not consider the problems of high groundwater levels and salinization, and they do not correspond to the well-established principle of engineering science „No irrigation without drainage", which has essential and binding implications for the design of institutions for large-scale irrigation systems.

The analysis, thus far, has shown that there are no simple solutions to adequately address high groundwater levels and salinization. The results obtained from the case study indicate that the effective control of high groundwater levels and salinization remains a difficult task for Water User Groups with the investments needed, both financially for adequate equipment, and in skills for mechanical, chemical and biological maintenance activities. Although more responsibility could be assigned to user groups, the nature of groundwater levels and salinization, and that of the required means, makes state regulation necessary. One working hypothesis of this study is that in large-scale jointly managed systems, although water user organizations can play an increasingly important role, the state, at least, continues to have a prominent function for the implementation of groundwater and salinization control means, particularly in investments in the project's drainage system. Whether the state continues to play an important monitoring and regulatory role with regard to the maintenance of the drainage system, or whether the state agency continues to be responsible for these services, depends on country-specific conditions and on local experiences. Whatever the options are in this respect, water user organizations must interact with government agencies. The specification of tasks for which the agency and the water user organizations are responsible remains an important issue, and arrangements for institutionalized cooperation are as important as incentive structures to instigate the implementation of control means. Issues of major importance are how the negatively affected farmers can express their demand for salinization control; how bargaining is structured, and how agreements can be enforced.

The case study indicates that there are two central issues for which solutions have to be elaborated. One refers to the maintenance of the drainage networks, and another deals with investments in on-farm drainage systems.

Maintenance of the project's drainage networks
Under the present transfer programme, the Water User Associations have no control over the publicly subsidized services, in particular the maintenance of the entire project drainage networks. This means that the financing of one of the most

important groundwater and salinization control measures is outside the control of the state agency, the water user associations, and the farmers. There are various solutions to this issue which are not only valid for the Turkish case but allow for generalization:

(First option) Under the present regulation, it would be advantageous if the user associations could conclude contracts with the state irrigation agency on the services needed with the necessary details, e.g., timing, quantity, quality, compensations for non-provision, etc. Such an agreement would give them power over the state agency to press for timely and adequate services.

(Second option) In the long-run, the state agency's services, based on contracts mentioned above, could be charged to all beneficiaries, whose fees would then form the budget at the state agency's discretion. It would then become necessary for the beneficiaries to exercise control over the state agency's expenses and the supplied services. The charges should be assessed by both the state agency and the WUAs, and in advance, because money loses considerable value due to inflation during the collection period. They would be collected by the WUAs. All farmers with irrigable lands should contribute, at least partly, towards maintenance costs even if they do not irrigate in one season, because the facilities always require maintenance.[5] Other users of the main drainage networks, i.e., the industries and the cities, should also contribute towards maintenance costs.

(Third option) In the particular case of Turkey, but not only, it could, however, be difficult to implement such a strategy because this would mean a general shift away from the highly centralized bureaucratic structure to locally independent, or, at least, partly autonomous units, with bureaucrats giving up their superior position and becoming accountable to the farmers. It seems more pragmatic that, after a period of transition, an elected upper-level user association takes over the responsibility. They could be established within each water sub-district where all user associations are members and represented by elected persons, and where the industries and the cities are represented. With the Water User Associations becoming fully responsible for operation activities, cooperation with the agricultural agency can be achieved at reasonable (viable) costs to reduce water inputs per crop and hectare.

Under all three options, the maintenance of at least the tertiary and secondary drains could be an issue for the WUAs, at least after a time for experience, with skilled labourers and adequate equipment being a prerequisite. But the maintenance responsibilites for the main drainage canals cannot be issued by the WUAs because a WUA's service area is designed around a main delivery channel which does not coincide with the drainage area of the agricultural effluents. If the WUAs

---

[5] In the particular case, modifications of the existing regulations are necessary: annual investment charges should be assessed on a real basis due to the high inflation rate. The legal regulation that the assessed O&M water charges can be reduced by political decisions, according to Law No.6200 (29), should be cancelled. Law No.6200 (Art.32) would have to be modified in order to accelerate the rate of collection of O&M water charges, and in regard to the services provided by the state agency. The present regulation, i.e., the one-off penalty of 10 percent, should be revised because it encourages late payments.

would also be responsible for the main drains which cross their service area, those WUAs located in the downstream areas would be confronted with higher maintenance costs, and, if these expenditures are neglected, with the negative external effects caused by the upstream irrigators as well. The main drainage systems can only be subject to a transfer if all WUAs are represented in an upper-level decision-making body which is authorized to make decisions on drainage basin issues. As long as this condition is not met, i.e., at present there is no institutionalized upper-level WUA unit that might be able to solve such problems and to mediate costs and benefits accrued from drainage, the state agency must continue to deal with this issue.

Investments in on-farm drainage systems and their maintenance

Investments in the on-farm drainage systems, and their maintenance are important issues where cooperation with the farmers is essential. The analyzed case, with its heavy delays in the execution of the on-farm works, has shown the weakness of the top-down approach, and the inadequacy of both investment strategies. A new investment strategy is required with active farmer participation, covering the whole process from project identification to the maintenance of the on-farm development works. A central element of this approach is that the government is sharing in the farmers' investments, and that the farmers take responsibility for financing their share. Instead of cost recovery, where the installations remain the property of the government together with the associated responsibilities, cost-sharing could guarantee that the farmers fully own the on-farm infrastructures (i.e., the completed facilities would be transfered to the farmers). This approach clearly states which part of the on-farm development investments is subsidized by the state, and which part is to be paid by the beneficiaries, including realistic interest rates. The farmers should commit themselves to sharing a substantial part of the on-farm development costs in real terms before any government support is given.[6] Priority should be given to organized farmers, as this would add to efficiency in terms of time spent by the state agencies' staff on technical assistance, and the administration of credits and repayments.

According to this investment approach for on-farm development works, the farmers, who become the owners of the on-farm installations, are responsible for their maintenance. The basic assumption is that farmers who participate in actual decision-making, and who share in the costs, will also have a positive attitude and feel responsible for the maintenance of the on-farm drainage infrastructures. There are, however, well-founded doubts that not all farmers are able, or willing, to invest: some farms are not large enough to obtain a sufficient net income for all family members; some families are no longer interested in agriculture because their members are employed in other sectors; other farmers lease fields or rent them out; under the various share-cropping arrangements, neither the tenants nor

---

[6] The requirement of cash for investments by the farmers should be abolished, or reduced, and the farmers should be offered a 100 percent financing package for the works through bank credits and grants. The farmers' share should be financed through the Agricultural Bank, with the Bank taking the responsibility for payments to farmers and the recovery of the farmers' share.

the owners have shown an interest in investments. If investments and maintenance depend on the individual farmers' decisions, they may also decide to save money through non-investment and non-maintenance, because the additional costs may exceed the anticipated benefits which derive from them. Because the negative impacts of non-maintenance are not only limited to those farm units directly concerned, i.e., the negative consequences will also affect others in the long-run, it seems to be necessary to limit the individual farmers' range of decision-making: the farmers would sign long-term user-right contracts with the state agency which clearly indicate their responsibilities, liability for damages, the maintenance instructions to be followed, the time intervals, the monitoring, etc. This option has the advantage that the farmers would commit themselves to these responsibilites with the agency exercising control over the proper and regular implementation of maintenance activities. Water deliveries could be linked to executed maintenance, or refused as a sanction if maintenance is not properly carried out.

The case study has shown some particularities which impose limits on the anticipation of effective management strategies, and the creation of proper institutional arrangements which can adequately address high groundwater levels and salinization. It can be concluded that highly dependent centralized decision-making settings cannot guarantee the implementation of effective groundwater and salinization control means. Important decisions, which are mentioned above, counteract to produce effective control measures. A challenge for future research would be whether autonomous, or semi-autonomous agencies are able to successfully deal with this issue, and the conditions therein, and how cooperation is institutionalized with what type of user groups participating. Another peculiarity of the case has been that the majority of the farmers were not negatively, or too severely, affected by high groundwater levels and salinization. It would be useful to empirically evaluate cases where the majority of farmers are heavily affected, together with the consequences for the farmers' participation.

Annexes

Annex 1
Data on the Service Area and the Irrigation and Drainage Networks

The Seyhan reservoir provides sufficient water for about 175,000 ha.[1] To date, the actual service area comprises 133,431 ha, or about 70% of the total project area, with 69,031 ha in the Tarsus Plain and 64,400 ha in the Yüregir Plain. After the construction of the drainage networks and the required pumping stations (Tarsus Plain 1,969 ha, Yüregir Plain 38,888 ha), an additional 40,657 ha will be able to be irrigated.

The delivery system in the Tarsus Plain comprises two conveyors; one, with a length of 54 km and a capacity of 20 m³/s, discharges water directly from the reservoir, and serves about 14,000 ha, and the Right Conveyor (40 km, 54 m³/s) which diverts water to 49,000 ha. There are 10 main channels feeding 740 km secondary and 725 km tertiary channels. Ten main drains, with a total length of 200 km, discharge into the rivers Berdan and Seyhan which then drain into the Mediterranean Sea; they are fed by 19 secondary (102 km) and 496 tertiary and quarternary drains (755 km).

The delivery system in the Yüregir Plain comprises 9 main channels, which are fed by the Left Conveyor (90 m³/s, 18.5 km), serving an area of 64,400 ha. Four main drains discharge directly into the Mediterranean Sea (east of the village Tuzla and near the village Karatas), and one drains into the Akyatan lagoon.

The capacity of the delivery system is designed to serve about 175,000 ha on a 24-hour-basis. Installations for flow regulation, e.g., gates and manual cross regulators, exist from the conveyors to the main channels, and from the main to the secondary system, but are lacking between the secondary and tertiary delivery systems. From the tertiaries or, sometimes, the quarternaries, water is withdrawn through siphons of various sizes at the farm outlets. In some areas, the wide spacing of the tertiaries still makes it necessary to cross other farmers' plots for irrigation. The delivery system is concrete lined and canalettes are used for the tertiaries and quarternaries.

---

[1] Between 1956 and 1986 the active storage capacity was reduced by 60 million cubic meters, which affected the flood control capacity, but not the capacity for irrigation and energy generation. It is estimated that approximately 12 million tons of sediment discharge annually into the reservoir, partly accumulating in the lake. The upstream Catalan Dam was therefore constructed with a high dead volume capacity, and the Cakit-Cayir Project implemented erosion control means in the upper catchment area of the Seyhan and its tributaries.

Annex 2
Irrigated Area as Percentage of Irrigable Land (1964-1989)[2]

Year	Irrigable area [ha]	Irrigated area [ha]	Percent
1964	18,030	18,727	103.9
1965	25,200	26,516	105.2
1966	41,512	40,145	96.7
1967	51,200	46,776	91.4
1968	57,322	52,129	90.9
1969	58,400	50,104	85.8
1970	58,400	36,929	63.2
1971	58,400	53,041	90.8
1972	58,400	58,199	99.7
1973	62,400	66,965	107.3
1974	83,550	82,517	98.8
1975	95,527	66,650	69.8
1976	98,547	45,022	45.7
1977	104,102	86,937	83.5
1978	110,480	73,399	66.4
1979	110,480	78,573	71.1
1980	103,000	84,670	82.2
1981	103,000	85,934	83.4
1982	103,000	92,575	89.9
1983	115,000	71,919	62.5
1984	119,000	95,756	80.5
1985	125,300	114,134	91.1
1986	133,431	116,198	87.8
1987	134,300	97,979	74.1
1988	134,300	102,032	77.1
1989	137,039	120,200	87.7

[2] Source: Tekinel 1992, p.10

Annex 3
Crop Patterns From 1966 to 1990 (Percentage Area)[3]

	Cereal	Alfalfa	Soybeans	Melons	Cotton	Maize	Paddy	Youngtree	Citrus	Vegetable	Others
planned	13.0	20.0	-	-	35.0	-	5.0	-	8.0	15.0	4.0
1966				0.1	97.4		2.4	0.1	0.1		0.1
1967				0.1	94.8		4.6	0.1	0.2	0.1	0.3
1968	0.2			0.1	96.4		1.7	0.8	0.5	0.1	0.3
1969	1.2			0.2	91.6		5.5	0.3	0.7	1.0	0.5
1970				0.6	88.7	1.1	4.9	1.2	1.0	0.4	2.0
1971				0.4	97.1	0.1	0.4	0.3	1.3	0.2	0.2
1972	7.0			0.3	90.0	0.1	0.8	0.9	0.9	0.2	0.2
1973	6.3			0.8	90.0	0.1	0.4	0.4	0.9	0.3	0.7
1974	0.6			0.6	96.5		0.4	0.4	1.0	0.3	0.1
1975				2.0	84.0	3.0	5.0	1.0	1.0	1.0	3.0
1976				4.0	81.0	3.0	4.0	1.0	3.0	1.0	2.0
1977				5.0	92.0			1.0	1.0		1.0
1978				9.0	81.0	1.0	4.0	1.0	2.0	1.0	1.0
1979	1.0			6.0	66.0	15.0	7.0	2.0	2.0	1.0	
1980				8.0	82.0	2.0	2.0	3.0	1.0		2.0
1981	1.0		1.0	10.0	77.0	1.0	3.0	4.0	2.0	1.0	
1982	16.0		6.0	12.0	48.0	5.0	4.0	6.0	2.0	1.0	
1983	0.8		8.4	10.4	62.4	1.6	2.7	7.1	4.1	0.7	0.2
1984	0.7		9.6	6.5	68.4	2.0	2.4	5.3	3.4	0.9	0.2
1985	2.6		16.6	8.5	51.6	8.7	4.8	2.1	2.4	1.0	2.1
1986	11.5		12.7	5.8	41.3	16.9	0.5	3.6	4.2	1.3	22.3
1987	1.0		19.0	9.0	37.0	20.0	1.0	3.0	6.0	2.0	3.0
1988	1.0		8.0	10.0	51.0	18.0	-	3.0	6.0	1.0	2.0
1989	16.0		9.0	6.0	35.0	23.0	1.0	3.0	4.0	1.0	15.0
1990	17.28		5.55	7.85	25.72	28.26	0.21	3.52	7.25	1.0	

[3]   Source: Tekinel and Kanber 1990, p.9

Annex 4
Data on Groundwater Levels and Dissolved Salt Concentrations[4]

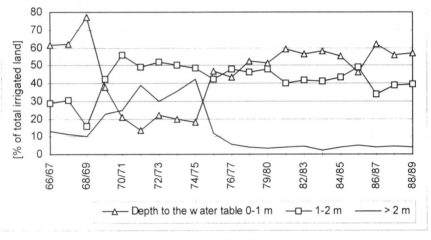

Fig. A.1. Critical maximum groundwater levels (1966-1991)

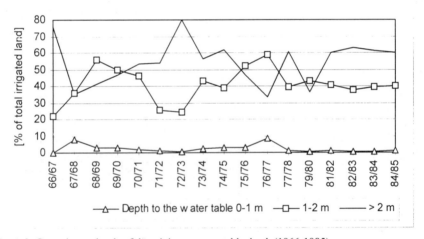

Fig. A.2. Groundwater levels of the minimum water table depth (1966-1985)

---

[4] Fig. A.1. to A.4. modified from FAO 1988, pp.16-17; 19-20

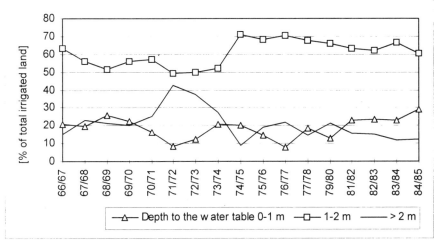

Fig. A.3. Groundwater levels during the most intensive irrigation months (1966-1992)

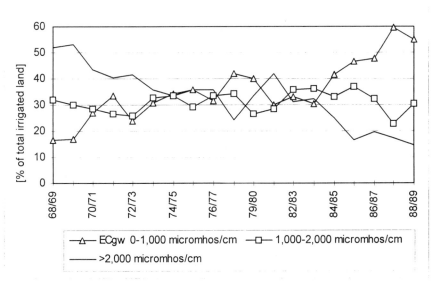

Fig. A.4. Dissolved salt concentrations of the groundwater during the months of irrigation (1968-1994)

Table A.1. Depth of groundwater levels in the peak irrigation months (1987-1993; in hectare)[5]

Year	0-1 m	1-2 m	2-3 m	3-4 m	>4 m
1987-88	7,901	17,290	3,227	128	-
1988-89	7,282	17,184	3,865	215	-
1989-90	6,714	18,300	3,442	90	-
1990-91	5,932	12,984	5,505	4,045	80
1991-92	4,631	7,655	12,802	3,378	80
1992-93	530	7,078	18,482	2,366	90

[5] Source: DSI, 6.Bölge Müdürlügü, ASO Subesi 1994 (Arca and Donma)

Table A.2. Areas with groundwater levels between 0-1m (1986-1992; in hectare)[6]

Years	Observation area [ha.]	Critical highest level				Critical lowest level				Peak irrigation months				Groundwater salinity Ec_{gw} > 5,000 micromhos/cm			
		Observation area [ha.]	[%]	Irrigation area [ha.]	[%]	Observation area [ha.]	[%]	Irrigation area [ha.]	[%]	Observation area [ha.]	[%]	Irrigation area [ha.]	[%]	Observation area [ha.]	[%]	Irrigation area [ha.]	[%]
1986-87	101,541	63,294	62.3	47,209	62.1	820	0.8	804	0.8	24,940	24.6	18,434	24.2	6,601	6.5	4,768	6.3
1987-88	106,571	56,784	53.2	51,097	53.1	349	0.3	318	0.3	23,145	21.7	20,898	21.7	5,021	4.7	4,497	4.6
1988-89	106,571	57,307	53.8	51,491	53.5	320	0.3	295	0.3	21,410	20.0	19,306	20.0	1,940	1.8	1,741	1.8
1989-90	128,571	66,712	51.8	59,905	51.9	340	0.3	310	0.3	21,012	16.3	18,766	16.3	2,466	1.9	2,201	1.9
1990-91	128,571	44,131	34.3	39,411	34.1	993	0.7	899	0.7	24,096	18.7	21,578	18.7	9,094	7.0	8,109	7.0
1991-92	128,571	61,317	47.7	54,480	47.5	575	0.4	515	0.4	22,311	17.3	19,896	17.2	4,799	3.7	4,333	3.7

[6] Source: DSI, 6.Bölge Müdürlügü, ASO Subesi 1993

Annex 5
The DSI's Responsibilities for Groundwater and Salinity Control, Observation Guidelines and Annual Reports

In the Lower Seyhan Plain, a groundwater table observation programme was started in 1952 by the Tarsus Irrigated Farming Research Institute which was attached to TOPRAKSU. In 1956, the Regional Directorate for State Hydraulic Works began its observation programme. In 1958, 78 observation wells had been installed, and in 1966, the DSI operated 605 wells. During the construction of on-farm works, most wells were destroyed, and they had to be renewed. In 1988, 880 groundwater observation wells had been installed. In general, one well is operated on every 100 ha. The responsibilities of the DSI's units for groundwater and salinity control programme are:[7]

Regional O&M division
* comparison of the annual groundwater tests of each project and writing of the annual report;
* control of the annual report, and programme implementation control;
* training of the technicians and the auxiliary staff;
* exchange of experiences and co-operation with other organizations.

Local O&M branches
* ensuring that the measurings are taken and to control them through the field operation units;
* monitoring of drainage and groundwater tests of other organizations, and possibly co-operate with them.

Field operation units in the water sub-districts
* to arrange measurings at each well in the last week of every month;
* maintaining the measuring stations, renewal of destroyed stations and setting up of new stations as required;
* preparation of the annual report;
* implementation and continuation of preventative work;
* training of personnel and monitoring of their work;
* determination of the sources of the groundwater recharge and establishing measures;
* implementation of the tasks arranged by the Regional Directorate.

Since 1966, the groundwater levels and the salinity content of groundwater have been observed regularly at monthly intervals, with the results being published in annual reports. The annual reports include precipitation data for the year; the irrigated area and the crop pattern; the water use efficiency; the water conveyance efficiency; data on groundwater levels, their salinity content, and their distribution

---

[7] The responsibilities are only documented in extracts. (DSI, Genel Müdürlügü, Isletme ve Bakim Dairesi Baskanligi 1987, pp.3-4)

within the investigated areas; conclusions and recommendations for preventive measures.[8]

An essential part of the annual reports are maps which demonstrate the spatial distribution of five different items:[9]

1. Map with the most critical groundwater levels
The map presents the highest position of the groundwater level of all observation wells in the course of a year. The cartographical presentation shows clearly the high groundwater level in irrgated areas. Groundwater levels between 0-2m indicate drainage problems.

2. Map with the lowest groundwater level
The map presents the lowest groundwater levels in a year. The presentation of the groundwater levels between 0-1m shows in which areas the groundwater stays at plant root level throughout the year. In these areas it is necessary to install on-farm drains.

3. Map of groundwater levels in the month of the maximum irrigation
The map evaluates the groundwater levels in the months of maximum irrigation. It allows the comparison with months of no irrigation und clearly shows the effects of irrigation.

4. Salinity content of groundwater
The map presents the distribution of the electrical conductivity of groundwater in the irrigation areas, and sets EC_{gw} 4,000 micromhos/cm as a critical level where the plant growth is going to be impaired.

5. Groundwater elevation map
The groundwater levels for each month are evaluated, and the groundwater surface topography is shown. For one year it indicates the highest and lowest level of the groundwater surface. For the operation of the system the crucial month is the one with the highest course, and therefore this is presented.

Annex 6
Annual Decree on Water Tariffs for O&M and Investment[10]

A. General Regulations
1. The costs for operation and maintenance are assessed annually, per hectare for each plant species, after the harvest. Should, during the vegetation season, two different crops be cultivated on one plot, the costs are calculated separately and the sum for both has to be paid. The costs for operation and maintenance are borne by the users, i.e., the owner, leaseholder or semi-leaseholder.
2. The operation and maintenance costs for the drainage systems are assessed per hectare, independently of the cultivated crops.

---

[8] Ibid., pp.13-14
[9] Ibid., pp.9-10
[10] T.C. Bayindirlik ve Iskan Bakanligi, Devlet Su Isleri Genel Müdürlügü: Yili Sulama ve Kurutma Tesisleri, Isletme-Bakim ve Yillik Yatirim Ücret Tarifeleri

3. If it is possible to measure the water volume in cubic metres the costs should be calculated and raised accordingly. The DSI is free to decide on how fees are raised.

4. Each farm shares in the operation and maintenance costs.

5. The operation and maintenance costs are based on the the size of the area. The smallest area is 0.01 ha. Smaller areas are not considered; areas greater than 50 square meters are treated as 0.01 ha.

6. If two different kinds of crops are cultivated on one area the tariff is deterrminded by the one with the highest water requirement. The areas are surveyed annually.

7. The tariffs are cheaper outside the irrigation season. The water required in autumn for germination is calculated at off-season rates. After that, the normal tariff is applicable. The term 'off-season' describes the time of the year not used for crop production but for soil improvement. Off-season areas are not surveyed.

8. If, during a year, new systems are put into operation the cost calculations are based on the costs of the running systems until the new tariffs are fixed.

9. If the water group takes over the control of the water distribution on a community level no tariff will be raised. If the control is withdrawn from the water group the DSI fixes a new tariff.

10. A contract is concluded between the DSI and the community setting out the tasks of the community. The system is then run by the group. If the community does not fulfill its tasks the DSI takes over the operation. The costs of the previous year including an annual increase rate serve as a guideline for the cost calculation.

11. If the land survey is carried out late, or not at all, the size given by the farmer is accepted. In that case, however, the permission of the ministry is required.

12. All farmers who want to irrigate have to forward their details to the DSI. They are not allowed to irrigate without having made an application. Should they, nonetheless, do so, fines are raised (10% of the operation and maintenance costs). The DSI can grant water or deny it if wrong information was given, or the formal procedures not kept.

13. The annual installment for investment costs for the irrigation and drainage system is fixed for each area (through resolution of the council of minsters). The tariffs are independent of the type or rotation of crops. Factories, mills, and brickworks are not charged.

14. If, within one year, more than one person farms one plot of land the investment costs have to be shared accordingly.

15. If the DSI cannot supply sufficient water the costs for the water delivered are raised using a guideline ([0.80]: e. g. 1.5 ha x 0.80 = 1.2 ha).

B. Price Reduction

16. If, during a year, two crops are cultivated on one parcel of land, and both are being irrigated, the normal tariff applies for the first crop, and a reduction of 50% is granted for the second.

17. If the systems are operated by the community or a cooperative, then the community or cooperative is entitled to a price reduction. The extent of the reduction is determined by the water volume, the local conditions, the crops, and their characteristics. A price reduction of 12% is granted if they participate in the operation, and a reduction of 16% if they participate in the operation and maintenance. Reductions can be up to 20%; depending on the type of system and the kind of service; reductions up to 25% can be given (with the agreement of the General Directorate) if they, additionally, participate in maintenance and repair works.

18. A price reduction of 25% is granted if water is used from other systems (pumps or private springs).

Annex 7
Contractual Agreements Between the DSI and the Water User Groups: The General Contract and Various Protocols[11]

1.
Contract of the Köy Tüzel Kisiligi's (KTK) Contribution to O&M Services

The contract partners, the name of the village , the size of the area to be irrigated und the irrigation channels are stated. The contract refers to repair and maintanance work as well as to the water distribution.

Duties and responsibilities

a) The responsible persons collect the demand notices of the water users and pass them on to the DSI.

b) To enable the DSI to prepare the daily water distribution and withdrawal schedule, the DSI has to be given the order in which the farmers register their daily demands; this system should prevent farmers from taking water when it is not their turn.

c) Clearing of the irrigation channels to the plots.

d) To provide assistance in settling disputes between water users, for example, over the access to water.

e) Destruction and polluting are to be prevented; damages have to be reported and the person concerned named.

f) Keeping up good relations between the DSI and the water users.

---

[11] The contract is a model agreement between the DSI and the Water User Groups, the protocols are examples.

g) Participation in the annual inspections, in which the necessary repair and maintenance works are assessed, and in which place they have to be carried out.

Operation
In order to carry out the above mentioned works, sufficently qualified staff have to be employed and the necessary technical equipment has to be provided for them. The personnel have to be renumerated accordingly, and the social security contributions have to be paid by the KTK. The KTK is liable for damages to third parties.

Discount arrangements for payments and accounts
The water tariffs are issued annually by resolution of the Council of Ministers, including the discount to be granted. The annual discount for the concerned village and the work to be done is laid down in a record. The KTK collects its share of the charges from the water users. The profit from the discount is registered in the village cash book. The village mainly uses this profit to pay for the agreed 0&M services; the rest can be used to pay for village development activities. All village expenditures paid out of this money must be confirmed with a receipt.

Relations to Third Parties
The rights-of-use are not transferable to Third Parties.

Cancellation
If the services are not carried out satisfactorily the contract can be unilaterally cancelled before the end of the period of notice. The water users cannot claim for damages.

Disputes
The court is responsible for settling disputes between the DSI and the KTK.

The contract is signed
in the name of the regional DSI administration, by the director of the 0&M division and, in the name of the village administration, by the village mayor and the village council.
The conclusion of the contract is laid down in Law No.6200, Min. 2,k.

2.
Protocols Which Fix the Price Reduction, the Rendering of Accounts, and for Maintenance and Repair Works

Protocol One
If the Water User Groups contribute to 0&M activities, the quality of their work is evaluated, which then is expressed in a percentage share of the total O&M costs. The protocol determines the share which the KTK receives in accordance with the published tariffs for O&M.

Protocol Two

defines the rendering of accounts. „As management of the KTK you have to collect and pass on the fees from the members, and you have to keep a record of income and expenditure in the following way:

1. maintain a list of group members;
2. issue a receipt for all income and expenditures;
3. register expenditures incl. documents and receipts;
4. expenditures for village services and irrigation services have to be registered separately;
5. income and expenditure are part of the annual general accounts which have to be recorded in the village cash-book;
6. the pages of the cash-book of the village administration have to be numbered throughout. Each page is checked by the administration and stamped with the note 'seen'.

All above listed tasks 1 to 6 have to be carried out before a set date. The accounts and the balance have to be presented to the head of the regional administration and the public prosecutor's office for control."

The details from paragraph 1 and 2 are publicly displayed on a notice board in the village coffee shop where they can be read by all water users. Protocols One and Two are signed by the village mayor, the village council and the representative of the Regional Directorate of the DSI.

Other protocols

includes precise instructions on how the works, e.g., painting of the metal parts of the weirs; sealing of the channels, have to be carried out. The DSI advises that the instructions have to be followed; in some cases, sanctions are mentioned, or, should the work not be completed, the discount will be reduced.

Annex 8
Data on the Water User Associations in the Project Area[12]

Water User Associations	Service area (ha)	No. of plots	No. of villages, municipalities, etc.	No. of water users	Main canal	General Assembly members
1. Cumhuriyet	1,765	1,277	3	795	YS0-8	17
2. Kuzey Yüregir	4,860	n.a.	12	n.a.	YS1	35
3. Akarsu	8,943 (7,662)	1,359	10	2,423	YS2	51
4. Cotlu	2,425	399	9	820	YS4	29
5. Güney Yüregir	16,890	n.a.	23	1,812	YS3	69
6. Gökova	4,289 (8,439)	n.a.	10	540	YS6	30
7. Yenigök	1,864	n.a.	7	n.a.	YS9	21
8. Kadiköy	9,808	1,571	19	913	YS8	57
9. Gazi	6,394	1,154	13	600	YS7	39
10. Seyhan	3,610	872	7	627	TS3/3TP	19
11. Toroslar	13,700	9,805	35	n.a.	TS1, TS2	56
12. Tarsus Onköy	11,983 (8,015)	2,790	10	1,667	TS9-TS10	36
13. Yesilova	3,590	680	8	427	TS3/1TP	20
14. Altinova	6,150	1,461	10	714	TS5-TS0-T1	27
15. Pamukova	12,037	4,863	18	2,618	TS6-TS7	51
16. Cukurova	ongoing					

[12] The entire service area of the stages I, II and III will come under the responsibility of Water User Associations. In 1994 and 1995, nine Water User Associations had been established in the Yüregir Plain, and six in the Tarsus Plain. The transfer contracts were signed in 1994 and 1995. The final data are from June 1995. (DSI, 6.Bölge Müdürlügü 1995b; DSI, Sixth Regional Directorate 1995)

Annex 9
Formation Statute of the North Yüregir Water User Association[13]

Paragraph 1. (Name) North Yüregir Water User Association
Paragraph 2. (Based) District of Adana-Incirlik
Paragraph 3. (Area of responsibility) Within the Province of Adana, the district of Yüregir, the areas including the villages of Alihocali, Camili, Dagci, Dedepinar, Köklüce, Suluca, Yürekli and the town councils of Incirlik, Yüregir, and Yakapinar, which are supplied by the main YSI canal.
Paragraph 4. The WUA is founded by the above named administrations, following the village act of 1580, paragraphs 133-148. Members of the WUA are all land owners, who benefit from the installation; they have to register and they are bound by the regulations.
Paragraph 5. All named administrations have the objective to take over, run, maintain, and repair the installations constructed and operated by the DSI, and to provide a service to the irrigation farmers; they follow the directions of the DSI for operation, maintenance and repair.
Paragraph 6. The WUA comprises of the General Assembly, the Commission, and the President.
Paragraph 7. (Formation of the General Assembly, its tasks, and its responsibilities) The beneficiary landowners of each administration (village administration, town council) elect two delegates for five years (and an appropriate number of reserve delegates); should a delegate, for whatever reasons, not belong to the adminstration unit anymore, he would lose his status both as delegate and member of the assembly.
Paragraph 8. Natural members of the assembly are the heads of the villages (Muhtars) and the mayors of the town councils; they have the same status as the elected delegates and participate in ballots and decision-making. The DSI has the status of observer.
Paragraph 9. (Formation of the General Assembly, its tasks, and its responsibilities)
1. Vote of the President, the deputy and the Commission.
2. The passing of the work schedule and work programme.
3. Discussion and decision of the budget; allocation of moneys for individual areas, approval of additional and special tasks.
4. Passing of the annual report of the previous year (preperation by the head of the assembly) with a two-thirds-majority.
5. Decision on staff appointment.
6. Decision on the rate of the contribution for the member-administrations for the following year's budget.
7. Reminders to defaulters (who for longer than a year did not fulfil their obligations).

[13] Kuzey Yüregir Sulama Birligi 1994

8. Decisions on property acquisitions.
9. Control of the employment of funds with regard to their purpose.
10. Discussion about the proposals of the head of assembly and resolution.
11. Approval of contracts with terms longer than 10 years and exceeding TL 10 million. (prices 1995).
12. Co-operation with public and international institutions.
13. Consideration of proposals of other WUA divisions and resolution.
14. Changes in the formation status.
15. Decision on the liquidation of the WUA.
16. Members of the assembly who did not participate in three meetings of the assembly cease to be members of the assembly (it is assumed that they have resigned).
17. Resolution on the minutes proposed by the head of the assembly.

Paragraph 10. (Meeting and quorum of the General Assembly) Meetings of the General Assembly take place in May and September. The General Assembly can be called together by the President or upon the request of one third of the members. Should the meeting not take place it can be convened seven days later by the President. In this case the number of the members is not important. The General Assembly has a quorum when at least 50% plus one member are present; decisions are reached with a simple majority.

Paragraph 11. (Legally valid resolutions) Resolutions of the General Assembly become legally valid after they have been signed by the provincial governor; they are passed to the provincial governments and local administration units for notification.

Paragraph 12. (Formation of the Commission, tasks and responsibilites) The Commission comprises of 7 members (head of the WUA, general secretary, accountant and 4 members elected by the delegates for one year). The general secretary must be an agricultural engineer.

Paragraph 13. (Tasks of the Commission)
1. First examination of the budget proposed by the President
2. Annual report (including the budget)
3. Decision on individual budget positions and alterations of allocations
4. Preparation of annual labour and investment programmes to be presented to the General Assembly
5. Audit of monthly account
6. Resolutions on increases or reductions of burdens and on the allocation of orders after reviewing cost and benefit
7. Decisions on renting and letting
8. Control of the accounts (whether monies were spent according to budget)
9. Accepting and implementing tasks in accordance with the law and regulations
10. Implementing the obligations laid down in the contract between the DSI and the WUA; referring to the following
    * operation, maintenance, and repair activities;
    * protection of the installations;
    * determination and charging of fines in the case of a breach of the above mentioned regulations and instructions by WUA members;

* determination of the water charges.

Paragraph 14. (Commission meetings and legally binding resolutions) Commission meetings take place at least every 15 days; it has a quorum with at least 50 % plus one of the members and resolutions are reached with a simple majority. In the event of a tied vote the chairman has the casting vote.

Paragraph 15. (Tasks and responsibilities of the President) The President is elected from amidst the delegates for 5 years. In the event that no candidate be found or elected, the heads of the village administration or town councils would take the chair in annual rotation.

1. He represents the WUA at administrations and at court.
2. He submits the budget draft, the annual report and the progress report, the annual plan and the annual programme to the authorities.
3. He is responsible for the implemantation of resolutions of the Commission and the General Assembly.
4. He is responsible for the budget or transfers the tasks to the General Secretary.
5. He ensures the collection of the takings and thereby pursues the interests of the WUA.
6. He convenes the meetings of the General Assembly.
7. He controls the annual programme.
8. He hears the Commission and refers demands, etc., to the General Assembly.
9. He can accept donations (without condition) on behalf of the WUA.
10.He administers the moveable and immoveable property of the WUA.

Paragraph 16. (Appointment of the personnel, tasks and authorities) The WUA employees are appointed by the President in accordance to the Civil Service Act, and other relevant laws for the province and town administration (gazette 23.6.1987, No.19496); the General Assembly decides on the appointments in its first sitting.

Paragraph 17. The General Secretary acts on behalf of the President; the President informs the General Secretary in writing about the assigned duties. The General Secretary is the head of the personnel.

Paragraph 18. (Income of the WUA)

1. The members pay an annual deposit, which is at least 1% of the previous year's budget.
2. The members make a one-off payment of TL 50,000/ha.
3. Donations
4. Water fees (operation, maintenance, repairs and related services for irrigation)
5. Financial aids and subsidies
6. Voluntary contributions/ shares
7. Fines
8. Sales, purchases, rents and interest
9. Miscellaneous sources

Paragraph 19. (Expenses of the WUA)

1. Expenses for hired services, and obligations which result from laws and regulations.
2. Expenses for administration and personnel.

3. Travel expenses for members of the General Assembly and the Commission, the President and the Deputy.
4. Expenses for debts, and expenses which arise from resolutions of the General Assembly or laws/ directives and contracts.
5. Costs for investments and projects.
6. Miscellaneous

Paragraph 20. (Budget of the WUA) In the budget decree, income and expenses of the WUA are estimated for one year; the General Assembly discusses the budget draft in September and decides on alterations. In the budget decree the members' contributions (defaults; installments) have to be registered by date. The decree contains, additionally, a price list for the services of the WUA.

Paragraph 21. The budget becomes legally binding with the signature of the Governor of the province of Adana. (...)

Paragraph 27. The WUA as a corporate body is responsible for all arrangements and securities. (...)

Paragraph 30. The General Assembly can revoke the membership of a delegate elected by a local adminstration in the event that the said member does not abide by the laws and regulations. All his rights and receivables are suspended for one year. The proceedings are initiated through the decision of the General Assembly.

Paragraph 31. The General Assembly can be dissolved by a resolution from a two-thirds majority. The dissolution is undertaken by the Commission and has to be finalised within one year.

Interim Regulations

Interim Paragraph 1. Not later than one month after the statutes come into effect the member adminstrations are called to a meeting chaired by the provincial Governor (or a person nominated by him); during this meeting the interim General Assembly, the interim President, and the interim members of the Commission are elected.

Interim Paragraph 2. The permanent bodies are elected 6 months after the beneficiaries have registered as members.

Interim Paragraph 3. The local adminstrations can themselves decide on the electoral procedure to be used to elect their delegates. Each local administration (village administration or town council) can either send an equal number of delegates, or the number can be determinded by the size of the land to be irrigated. It is, however, compulsory, to send at least two delegates and two deputy delegates.

Annex 10
Contract Between the General Directorate for State Hydraulic Works and the North Yüregir Water User Association[14]

Paragraph 1
The contract concluded between the DSI and the beneficiaries, who are repre-sented by the North Yüregir Water User Association, is based on the Legal Act No.6200, Paragraph 2 (K).

Paragraph 2
On conclusion of the contract, the operation and the maintenance of the system is taken over by the beneficiaries, who are the residents in the areas that come under the province of Adana, the district Yüregir, the town councils, and the village adminstrations, and who are supplied by the system YS1 built by the DSI.

Paragraph 3
The beneficiaries take over the operation of the system together with the related tasks (the abstraction of water from the Seyhan reservoir to the field outlets). The appropriate organizational structure is set up to secure the service.

Paragraph 4
The DSI and the beneficiaries organization write an annual inspection report; in addition, the tasks to be carried out by the beneficiaries themselves, or with exter-nal support, are laid down. The tasks are strictly and precisely carried out, and if not, the beneficiaries have to bear the incurred costs in full. The annual inspection report and the DSI-control guidelines and regulations are followed. The beneficiar-ies take over the safeguarding of the system, with the necessary maintenance and repair works.

Paragraph 5
The operation and maintenance of the installations is carried out according to the directives of the DSI.

Paragraph 6
In the event that the operation, maintenance and repair work is not carried out according to the directives and damages occur, the beneficiaries have to bear the costs.

Paragraph 7
The WUA has to assess the costs for maintenance and repair, and have the amount, approved by the DSI as sufficient, available. The DSI controls the expenses, and

---

[14] Kuzey Yüregir Sulama Birligi 1994

the user organization has to follow the DSI's directives for maintenance and repair. The users bear the costs for maintenance and repair in full.

Paragraph 8
The fees charged to each member for operation, maintenance and repair have to be recorded on an account sheet by the user organization and presented to the DSI for confirmation by April at the latest. Afterwards the fees can be collected.

Paragraph 9
The WUA is obliged to provide all information required by the DSI.

Paragraph 10
The investment costs are borne by the beneficiaries (according to Act No.6200 and other administration regulations).

Paragraph 11
The rights laid down in this agreement cannot be transferred to a third person; a written consent by the DSI is required, should such a transfer be requested. The use of the system is only permitted under the agreed conditions.

Paragraph 12
The WUA is fully responsible for the staff employed by or working for them. It has to be ensured that social insurance agreements and working conditions are satisfactory.

Paragraph 13
No alterations to the systems, regarding the construction or other, can be carried out without written permission from the DSI.

Paragraph 14
In the event that the user organization does not keep to the terms of the contract, even after a written warning, the DSI has the right to abolish the contract. In the event that the contract is cancelled there is no entitlement to compensation for damages. The system has to be returned to the DSI in its initial condition. In the event that the contract is cancelled by mutual consent, the user organisation is responsible towards third persons.

Paragraph 15
Quarrels are settled before the Court of Adana.

Paragraph 16
The contract signed on 21.3.1994 comes into force after confirmation by the Ministry of Public Works and Settlement.

Signed
on behalf of the General Directorate for State Hydraulic Works by the Sixth Regional Directorate; on behalf of the beneficiaries by the North Yüregir Water User Association.

Annex 11
The General Directorate for Rural Services' Experiences with Groundwater Cooperatives[15]

The construction of groundwater irrigation projects is under the joint responsibility of both the DSI and GDRS. After the repayment of the DSI's capital costs, the infrastructure becomes the property of the cooperatives; the GDRS investment costs are free of charge. Before the construction of the irrigation project, GDRS establishes water user cooperatives. The cooperatives are directly accountable for all recurrent O&M expenses. The process of establishing a cooperative is initiated by a farmer or a group of farmers in a village, one of whom is appointed to petition the GDRS for its services. The petition is sent to the DSI for proofing (soil survey, groundwater resources). The GDRS then informs the petitioner that he must organize fifteen people to form a cooperative (before 1992, only seven people were necessary). The minimum deposit necessary to open a local bank account is equivalent to the start-up fee for initiating the cooperative. The founders must forward this money from their own resources.

The original founders elect from among themselves the first Cooperative Board of Directors who consists of five non-salaried partners including the President, the Vice-President, a Treasurer, and two representatives. This Board decides on the membership fee, assuming that 100% of those who will benefit from the project will be committed to the cooperative. New members can be registered, and 25% of the membership fee is collected immediately, with the remaining 75% to be paid within three years, without interest. A General Assembly of the cooperative is established where all members have equal rights; each farmer with title deeds on irrigated farmland has one vote. The advantages of being members and joining the cooperative are: the cooperative receives a subsidized electricity rate which is approxiamtely 35% of the normal household electricity rate; the cooperative determines the order of water availability. Non-members can receive water but they will be last in the line of recipients, and they are not eligible for the electricity subsidies. The General Assembly meets annually before the irrigation season to elect the Board of Directors. Elections require a turnout of 50%, plus one member of the General Assembly. If this turnout is not realized, the elections are postponed for seven days. At the second meeting of the General Assembly, Board elections are held regardless of the number of participants, which, if there is no major conflict of interest, averages between 10-25%. As of 1992, Cooperative by-laws were amended with a clause stating that a partner who does not attend the General Assembly annual meetings for three consecutive years will be dismissed from the

---

[15] See The World Bank 1992, Annex 8

cooperative. As previously mentioned, each farmer has one vote, but a farmer with a large estate can legally divide the estate amongst his family and obtain a voting advantage (for their opinions). Tenant farmers are considered associate partners and may attend General Assembly meetings in voice, but, however, they are not eligible to vote. Absentee voting is not permitted, and votes are cast by secret ballot.

Apart from operation and maintenance accountability to the water-users, and in order to function effectively, the cooperatives have also established means to re-solve conflict. Most problems are presented to the Board of Directors, and if they cannot be settled at this level, a plaintiff may petition for a meeting of the General Assembly. A meeting is usually held within fifteen days, attended by those who wish to vote on the issue. As a last resort, if the matter is not resolved, the issue is taken to the local court system. The cooperative employs from between six to ten members for its operational staff. During the irrigation season, these employees are on duty seven days a week, 24 hours a day. All salaries are paid from the coopera-tive budgetary resources, which are collected through annual fees. By general consensus, the partners prefer to employ temporary labourers to attend to mainte-nance needs before the irrigation season rather than assuming this responsibility themselves. The cooperative hires from between one to fifteen people for a period of three to fifteen days to de-silt, de-weed and repair canals. Additionally, the cooperative hires an external accountant to finalize the budget at the end of each month and to assist the Treasurer in preparing the irrigation invoices for each wa-ter user. The billing procedures are not determined by by-laws. Each cooperative reserves autonomy in determining water user fees, salaries and investment ven-tures. Regulating the irrigation rotation and determining the appropriate water user fee is the fundamental obligation of the cooperative, and each individual coopera-tive advances its unique system of irrigation rotation and water use accounting. Irrigation systems supporting homogeneous crops often adopt a lottery system for water use rotation. On a pre-announced registration day, each irrigator pays a small registration fee and is assigned a call number for the most accessible well on a first-come, first-served basis. Once the irrigation season commences, the rotation order from each well is determined from the numbers randomly picked out of all the registered numbers; a call number does not re-enter the lottery until all the remaining numbers have been called.

The water use duration is documented by the well operator; the water fee is thus determined by the power consumption of the well in use. The lottery is drawn at the same time every evening during the entire irrigation season at the town square. Irrigation schemes with heterogeneous cropping patterns adopt a more complex variation of this lottery system which takes into consideration the different crop water requirements. Regardless of the method, the order of water use is rotated every year. The major share of the 0&M costs is the power bill which the coopera-tive is directly accountable for. The first bill is submitted three months after the beginning of the irrigation and presents no problem, but the second covers the months of high water consumption for crops. Here arrears occur, which are tempo-rarily covered by the cooperative or its wealthier members. A penalty of 10% per month is levied for late payment of water charges, but conflicts are usually resolved without recourse to sanctions. Additional expenditures, such as invest-ments, are debated during the annual General assembly meeting before the irriga-tion season. One cooperative invested, for example, in time meters to incorporate an honour-system of water use so that each farmer could assume personal respon-sibility for operating the well used for his irrigation period. Each farmer communi-cates with the person next in line for irrigation to arrange a transfer of responsibil-ity. They meet at the well where the first farmer records in a log book the meter reading when he stops irrigating and the next farmer records the meter reading when he starts irrigating, thereby establishing an internal system of check and balance.

Annex 12
Water Rights in the Majelle, the Ottoman Codification of Moslem Law (Shari'a)[16]

The basic principle in the Majelle with regard to water states that „water, grass and fire are free to be used by all. In these three things mankind are partners".[17] This principle regards water as a gift of Nature (or as a religious charity) in the use of which every person has an equal right whether one is moslem or not. It objects to tribal factionism emphazising equality and fraternity.[18] Water is thus not subject to ownership rights but to community or public use, and as such, water needs no legal definition. This definition, i.e., water is „free for the use of all", has often been misinterpreted leading to the assumption that water could not become a commod-ity, and that there would be no property or user-rights definition, and in particular, that water could neither be sold nor transferred. Wohlwend states that „in all tradi-tional water law systems, a distinction is, however, made between water in motion, or present in a non-measurable quantity, and water separated from its natural source and contained in measurable receptacles. In this case water (...) having become quantifiable, acquires the status of a legal thing and qualifies for private appropriation. From a natural resource, water has, however, become a commodity, and he who owns it enjoys, by definition, an exclusive right of use thereon".[19] The quantifiable and separable commodity water is subject to ownership rights. In this

---

[16] Between 1854 and 1876 the Ottoman Civil Code, the Majelle, was promulgated sepa-rating religious and secular affairs. The Majelle was framed on the basis of the Napo-leonic Code by Ottoman legislators who selected from its rules those in conformity with, or upholding Moslem principles. The Majelle, first published in Turkish, was later translated into Arabic and put into force throughout the Ottoman Empire. But, al-though the Majelle was the unique codification of Moslem law (Shari'a), there were three islamic schools each with more than one law school differing in the interpreta-tion of Shari'a. In Turkey, for example, the Sunnites recognized four schools of law of which the most important was the Hanifite. Furthermore, when customary law was not contradicting with the Shari'a, the Majelle did not always, and everywhere, replace it. (see Caponera 1973)

[17] Majelle (undated), Art.1234

[18] Caponera 1973, p.10

[19] Wohlwend 1981, p.4

sense, the Majelle defines different kinds of property and user-rights on surface water (and groundwater), restrictions on its use, and modalities for its transfer.

Public property, private property, common property
With regard to surface water resources, the Majelle differenciates between (1) 'mulk' property which is the private property of an individual, (2) common (or collective) property owned by shareholders, (3) 'public' property in the sense of 'unowned' where everyone has free access, and (4) state property. Private property and common propery are at the disposal of the owners or co-owners, but the rights are limited with respect to the rights of affected persons: the owners and co-owners are free to use it unless the rights of others are excessively damaged, that is if the original benefit becomes distorted.[20]

Property rights on rivers
Water in the rivers is subject to public, i.e., unowned, and to private property rights. Public rivers are rivers which are not divided, and 'divided' means that they do not enter into channels which are the private property of a number of persons (examples of public rivers are the Euphrates and the Nile). Rivers, which enter into channels, can be owned as private property, and these channels are owned in shares. Two kinds of 'owned rivers' are distinguished. The first type refers to rivers which are distributed and divided among shareholders. Only if these rivers flow on unowned land, or if their water is not entirely used by the owner of this land, is the surplus free to be used by the public. The second type refers to private rivers whose water is spread and divided over the land of a limited number of persons and used within the boundaries of their land.[21]

Acquisition of private ownership and transferability
Private (or 'mulk') ownership can be acquired in different ways: by transfer from one owner to another as a gift; by inheritance; by occupation of a good which was free for the use of the public.[22] Article 216 defines that „it is lawful to sell right of passage (‚i.e., navigation),[23] a right to take from running water,[24] the right of overflow,[25] and to sell water „as subject to the water pipes". For commonly owned property rights, the Majelle defines that one may sell his undivided share without obtaining permission of his co-owner.[26]

---

[20] Majelle, Art. 1192; 1197; 1199
[21] The Majelle does not define how conflicts between upstream and downstream riparians are settled and mediated.
[22] „For example: When someone has taken water from a river with a vessel like a jug or a cask that water being kept and stored in that vessel, becomes the property of that person. No other than he has the right to use it." (Majelle, Art. 1249)
[23] Majelle, Art. 142 and 1224
[24] Ibid., Art. 143, 955 and 1262
[25] Ibid., Art. 144 and 1224
[26] Ibid., Art. 139 and 215

Prior appropriation

The Majelle recognizes a kind of prior appropriation system for property and user-rights where senior rights are superior over junior rights. Rights acquired since immemorial times (kadeem)[27] are attributed the same degree of priority if they do not interfere/conflict with the Shari'a Law. The rights of way, „rights of having a water channel and the right to discharge water (...) which is from time immemorial observed (...) are left and continued according to the way they have come from time immemorial" as long as there is no proof to the contrary.[28]

Priority of rights

The Majelle defines priority of rights for different purposes, e.g., the right of thirst, the right to irrigate and to water animals, and the right to operate mills. The right to drink water is compared with the use of air and (sun)light, and regarded as un-restricted.

Irrigation rights

With respect to irrigation „every one can irrigate his fields from rivers which are not owned by anyone (i.e. public rivers) and can open a canal or water channel to irrigate his fields".[29] But one may not cut off a river entirely nor, as a generel condition, should he damage or interfere in the rights of others. The right of water-ing the fields and animals from privately owned rivers belong to the owner. No person can use the water for irrigation purposes without the permission of the owner (but others may use it for drinking).[30] If rivers are commonly held, one can-not open a channel or conduit without the permission of the co-owners.[31] The banks of water channels crossing the site of another person belongs to the owner of the channel.[32]

Rights of way

Rights of way are to be guaranteed, and owners whose land is passed by a water channel cannot refuse the right of way, and also for repair and cleaning activities. Only if the owner granted the initial permission, has he the right to retract it.[33]

Repair and maintenance regulations

Regulations are specified for the repair and maintenance. The co-owners have to repair the property held in common proportionately to their shares. The upkeep of a river which is commonly owned falls on its owners, i.e., on those who have rights to appropriate water for irrigation and for watering animals.[34] If one co-owner undertakes the repair with the leave of the other co-owners, or, in the case of their

---

[27] 'Kadeem' means the origin not known by anyone.
[28] Ibid., Art. 1224
[29] Ibid,, Art. 1265
[30] Ibid., Art. 1267
[31] Ibid., Art. 1269
[32] Ibid., Art. 1290
[33] Ibid., Art. 1228
[34] Ibid., Art. 1322

absence with the leave of a judge, the other co-owners have to recover the expenditure/expenses. The Majelle differentiates divisible and indivisible common property; if division is possible and done by order of a judge, the person can do what he likes with his own share (maintain it or not). If some co-owners decide on the maintenance of a river and others object, the regulation differs between public rivers, where users have collective user-rights, and privately owned rivers. Public rivers are maintained by common actions of all co-users and objectors are compelled to comply. If all co-users object, they can be forced to take part in maintenance. Those who refuse to clean private rivers have to pay a portion of the expenses; non-payers are prohibited from taking their benefit from the river, i.e., their share of water. The repair and cleansing of commonly held rivers begins upstream, and all shareholders take part from the beginning and must contribute to the initial expenses. When upstream riparians have met their maintenance obligations, they are no longer bound to contribute to the maintenance of the downstream section. The shareholders are free from obligations when their land is passed. The most downstream shareholder also shares in the expenses.

References

Abdellaoui, R.M. 1987: Small-scale Irrigation Systems in Morocco: Present Status and Some Research Issues. In: International Irrigation Management Institute (IIMI) and Water and Energy Commission Secreteriat (WECS) of the Ministry of Water Resources, Government of Nepal 1989: Public Intervention in Farmer-Managed Irrigation Systems, pp.165-173. Sri Lanka

Abernethy, Charles L.; Kijne, Jacob W. 1993: Managing the Interactions of Irrigation Systems with their Environments. In: DVWK Bulletin 19, pp.75-105. Hamburg, Berlin

Achtnich, Wolfram 1980: Bewässerungslandbau. Agrotechnische Grundlagen der Bewässerungswirtschaft (Irrigated agriculture, and the principles of agricultural technology). Stuttgart

Adana Tarim Il Müdürlügü Kayitlari (cesitli yillar) (several years). Adana

Adana ve Karatas meteoroloji istasyonlarinda yagis, sicaklik ve buharlasma (Precipitation, temperature and evaporation at the meteorological stations Adana and Karatas). Adana

Ahmad, Afroz; Singh, P.P. 1991: Environmental Impact Assessment for Sustainable Development: Chittaurgarh Irrigation Project in Outer Himalayas. In: Ambio Vol.20 No.7, pp.298-302

Akalin, Hasan Tahsin 1966: Probleme der Neuordnung des Boden-, Wasser- und Forstrechts in der Türkischen Republik unter Berücksichtigung der landwirtschaftlichen Entwicklung (The reorganization of land, water and forestry rights in the Turkish Republic considering agricultural development). Freiburg im Breisgau

Aktas, Yasar 1976: Landwirtschaftliche Beratung in einem Bewässerungsprojekt der Südtürkei. Organisation und Arbeitsweise (Agricultural extension in an irrigation project in South Turkey. Organization and operation). Sozialwissenschaftliche Schriften zur Agrarentwicklung, Band 18. Saarbrücken

Alchian, Armen; Demsetz, Harold 1973: The property rights paradigm. Journal of Economic History, 33, pp.16-27

Arrow, Kenneth J. 1951: Social choice and individual values. New York, Wiley&Sons

Baden, John; Stroup, Richard L. 1981: Bureaucracy vs. Environment. The Environmental Costs of Bureaucratic Governance. The University of Michigan Press

Balaban, Ali 1972: Seyhan Ovasi Sulamasi ve Sorunlari (Irrigation in the Seyhan Plain, and its problems). Ankara

Bardhan, Pranab 1993: Analytics of the Institutions of Informal Cooperation in Rural Development, in: World Development, Vol.21, No.4, pp.633-639. Pergamon Press UK

Barghouti, Shawki; Le Moigne, Guy 1991: Bewässerung und umweltpolitische Herausforderung (Irrigation, and its ecopolitical challenges). Finanzierung & Entwicklung, Juni

Barrow, Chris 1987: Water resources and agricultural development in the tropics. New York

Barton, R.F. 1969: Ifugao Law. Berkeley and California

Baumann et al. 1984: Ökologische Auswirkungen von Staudammvorhaben (Ecological impacts of dams). München, Köln, London

Benoir, Daniel; Harrison, James 1976: Raising Farmers Incomes Through A Professional Extension Service. Welthungerhilfe, Bonn

Bergmann, Hellmuth; Boussard, Jean-Marc 1976: Guide to the economic evaluation of irrigation projects

Berk, Metin 1980: Public Policies Affecting the Distribution of Income Among Cotton Producers in Turkey, in: Özbudun and Ulusan (eds.): The Political Economy of Income Distribution in Turkey. Holmes & Meier Publishers, London

Berkes, Fikret (ed.) 1989: Common Property Resources. Ecology and community-based sustainable development. Belhaven Press, London

Bernholz, Peter; Beyer, Friedrich 1984: Grundlagen der Politischen Ökonomie (Principles of Political Economics). J.C.B.Mohr (Paul Siebeck) Tübingen

Betke, Dirk 1989: Die Umweltfrage (The environmental problem), in: Chinas Wirtschaft zu Beginn der 90er Jahre, pp.54-82. Hamburg

Bilen, Özden; Uskay, Savas 1992: Comprehensive Water Resources Management: An Analysis of Turkish Experience. In: LeMoigne et al.(eds.), Country Experiences with Water Resources Management. Economic, Institutional, Technological and Environmental Issues. World Bank Technical Paper No.175, pp.143-150. Washington, D.C.

Biswas, Asit K. 1985: Evaluating irrigation's impact: guidelines for project monitoring. Ceres (FAO), No.106

Blaikie, Piers; Brookfield, Harold 1987: The Degradation of Common Property Resources, in: Blaikie and Brookfield: Land Degradation and Society. London, New York

Boguslawski, Michael von 1986: The crop production of Somalia. In: Conze and Labahn (eds.) 1986: Somalia. Agriculture in the Winds of Change, pp.23-54. epi Dokumentation 2. Saarbrücken-Schafbrücke

Bos, M.G.; Nugteren, J. 1974: On Irrigation Efficiencies (ILRI Publication No.19). Wageningen (The Netherlands)

Bottrall, Anthony F. 1981: Comparative Study of the Management and Organization of Irrigation Projects. World Bank Staff Working Paper No.458. Washington, D.C.

Boyko, Hugo 1966: Salinity and Aridity. New Approaches to Old Problems. The Hague

Bradhan, Pranab 1989: The New Institutional Economics and Development Theory: A Brief Critical Assessment. World Development, Vol.17, No.9, pp.1389-1395. Pergamon Press UK

Breton, Albert; Wintrobe, Ronald 1982: The logic of bureaucratic conduct. Cambridge University Press

Bromley, Daniel W. 1982a: Improving Irrigated Agriculture. Institutional Reform and the Small Farmer. World Bank Staff Working Papers, No.531. Washington, D.C.

Bromley, Daniel W. 1982b: Land and Water Problems: An Institutional Perspective. American Journal of Agricultural Economics, Vol.64, pp.834-844

Bromley, Daniel W. 1989: Economic Interests and Institutions. The Conceptual Foundations of Public Policy. Basil Blackwell, New York

Bromley, Daniel W. 1990: Arresting Renewable Resource Degradation in the Third World: Discussion. American Journal of Agricultural Economics, Vol.72, pp.1274-75

Bromley, Daniel W.; Taylor, Donald C.; Parker, Donald E. 1980: Water Reform and Economic Development: Institutional Aspects of Water Management in the Developing Countries. Economic Development and Cultural Change, Vol.28, No.2, pp.365-387. Chicago

Buchanan, J.M. 1979: Eine ökonomische Betrachtung der 'wissenschaftlichen Politik' (An economic view on scientific policy), in: Pommerehne and Frey (eds.): Ökonomische Theorie der Politik. Springer Berlin, Heidelberg

Buchanan, James M. 1968: The Demand and Supply of Public Goods. Rand McNally Chicago

Buchanan, James M. 1978: From Private Preferences to Public Philosophy: The Development of Public Choice. In: Buchanan; Rowley; Breton; Wiseman; Frey; Peacock: The Economics of Politics. The Institute of Economic Affairs

Buchanan, James M.; Tullock, Gordon 1962: The Calculus of Consent. Logical foundation of constitutional democracy. Ann Arbor, University of Michigan Press

Bülbül, Mehmet 1973: Adana Ovasi Tarim Isletmelerinin Ekonomik Yapisi, Finansman ve Kredi Sorunlari (Farm economy, investment and credit problems in the Adana Plain). Ankara

Cabanilla, L.S. 1984: Study of Operation and Maintenance Problems in Irrigation: The Philippines Case. Manila, The Philippines, USAID

Caponera, D.A. 1973: Water Laws in Moslem Countries. FAO Irrigation and Drainage Paper No.20/1. Rome

Carapetis; Levy; Wolden 1991: Sub-Saharan African Transport Program: The Road Maintenance Initiative. Vol.II, Readings and Case Studies. EDI Seminar Series. World Bank USA

Carruthers, Ian D. (ed.) 1983: Aid for the Development of Irrigation. Organisation for Economic Co-Operation and Development. Paris

Carruthers, Ian D. 1993: Water Management and the Environment. In: DVWK Bulletin 19, pp.7-19. Hamburg, Berlin

Carruthers, Ian D.; Clark, Collin 1981: The Economics of Irrigation. Liverpool University Press, Liverpool

Carruthers, Ian D.; Morrison, Jamie 1994: Maintenance in Irrigation. A Review of Strategic Issues. Eschborn

Caswell; Lichtenberg; Zilberman 1990: The Effects of Pricing Policies on Water Conservation and Drainage. American Journal of Agricultural Economics, Vol.72, pp.883-890

Cernea, Michael M.; Meinzen-Dick, Ruth 1992: Design for Water User Associations: Organizational Characteristics. In: LeMoigne, Barghouti and Garbus (eds.), Developing and Improving Irrigation and Drainage Systems. World Bank Technical Paper No. 178. Washington, D.C.

Chambers, Robert 1988: Managing Canal Irrigation. Practical Analysis from South Asia. Cambridge University Press

Ciriacy-Wantrup, S.V. 1975: 'Common property' as a concept in natural resources policy. Natural Resources Journal, Vol.15, No.4, pp.99-113

Coase, Ronald 1937: The Nature of the Firm. Economica 16, pp.386-405

Coase, Ronald 1960: The Problem of Social Cost. Journal of Law and Economics, 3, pp.1-44

Consortium TAHAL-ECI 1966: Seyhan Irrigation Project. Feasibility Report Stage III. Denver, Tel Aviv

Consortium TAHAL-ECI-SUIS 1968: Seyhan Sulama Projesi Planlama Raporu 3TP-2YP, Safha II (Planning report for the Seyhan Irrigation Project 3 TP-2YP, Stage II). July 1968. Adana, Ankara, Denver, Tel Aviv

Consortium TAHAL-ECI-SUIS 1970: Seyhan Irrigation Project. Extension Service. Stage I Area 1969. Programme Activities. Adana, Ankara, Denver, Tel Aviv

Consortium TAHAL-ECI-SUIS 1971: Seyhan Irrigation Project, Stage II. Analysis and Proposals for Irrigation Water Charges for the Seyhan Irrigation Project. Adana, Ankara, Denver, Tel Aviv

Consortium TAHAL-ECI-SUIS 1973: Seyhan Irrigation Project, Stage II. Operation and Maintenance Annual Report, March 1.1972 to February 28.1973. Adana, Ankara, Denver, Tel Aviv

Cruz, Ma. Concepcion J. 1989: Water as Common Property: The Case of Irrigation Water Rights in the Philippines. In: Berkes, Fikret (ed.) 1989: Common Property Resources. Ecology and community-based sustainable development. Belhaven Press, London

Cummings, Ronald G.; Nercissiantz, Vahram 1992: The Use of Water Pricing as a Means for Enhancing Water Use Efficiency in Irrigation: Case Studies in Mexico and the United States. Natural Resources Journal, Vol.32, No.4, pp.731-755

Dabag, Mihran 1988: Das Verwaltungssystem der Türkei. Ein Abriß (The administrative system in Turkey. An outline), in: Franz Ronneberger (ed.), Zwischen Zentralisierung und Selbstverwaltung. Bürokratische Systeme in Südosteuropa. Südosteuropa Jahrbuch, 18.Band. München

Dani, Anis A.; Siddiqi, Najma 1987: Institutional Innovations in Irrigation Management: A Case Study from Northern Pakistan. In: International Irrigation Management Institute (IIMI) and Water and Energy Commission Secreteriat (WECS) of the Ministry of Water Resources, Government of Nepal: Public Intervention in Farmer-Managed Irrigation Systems, pp.71-89. Sri Lanka

de Horter, Harry; Zilberman, David 1990: On the Political Economy of Public Good Inputs in Agriculture. American Journal of Agricultural Economics, Vol.72, pp.131-137

de Janvry; Sadoulet; Thorbecke 1993: Introduction. World Development, Vol.21, No.4., pp.565-575. Pergamon Press

Deutsche Gesellschaft für Technische Zusammenarbeit (GTZ) 1993: Irrigation and the Environment. A Review of Environmental Issues. Part I. Eschborn

Deutscher Verband für Wasserwirtschaft und Kulturbau e.V. (DVWK) 1983: Fachwörterbuch für Bewässerung und Entwässerung (Dictionary for Irrigation and Drainage). Bonn

Dhillon, G.S.; Singh, T.; Kumar, M. 1981: Stability of the Irrigation System in the Indus Basin. Water International, 6(1981), pp.83-91

Diestel, Heiko 1982: The costs and the worthiness of efforts to identify, to monitor and to control salt balances in irrigation schemes. ICID, 13th Regional European Conference, Lissabon, Paper 24. pp.13

Diestel, Heiko 1987: Water management to avoid salinization. ICID, 13th Congress, Rabat: Q.40, R.76, pp.1165-1176

Diestel, Heiko 1993: Reactions of Water Management to the Salinity of Soil and Water. In: DVWK Bulletin 19, pp.187-199. Hamburg, Berlin

Dinc, Oguz A. 1988: Asagi Seyhan Ovasi Tuzlu Topraklarinin Landsat - 5 TM Sayisal Uydu Görüntülerinden Yararlanarak Incelenmesi Üzerinde Arastirmalar (Advantages of Landsat for research on soil salinity in the Lower Seyhan Plain). Adana

Dinc, Ural; Sari, Mustafa; Senol, Suat; Kapur, Selim; Sayin, Mahmut; Derici, Rifat; Cavusgil, Veysel; Gök, Mustafa; Aydin, Mehmet; Ekinci, Hüseyin; Agca, Necat; Schlichting, E. 1990: Cukurova Bölgesi Topraklari (Soils in the Cukurova Region). Adana, Stuttgart

Dinc, Ural; Senol, Suat; Kapur, Selim; Sari, Mustafa; Derici, Rifat; Sayin, Mahmut; Cavusgil, Veysel 1986: Formation, Distribution and Chemical Properties of Saline and Alkaline Soils of the Cukurova Region. Adana

Downs, Anthony 1967: Inside Bureaucracy. Little, Brown and Company Boston

Downs, Anthony 1968: Ökonomische Theorie der Demokratie (Economic theory of democracy). J.C.B.Mohr (Paul Siebeck) Tübingen

Drenaj Etüdleri 1957. Cukurova'da Seyhan Barajindan Sulanacak Yüregir Ovasinda Drenaj Ihtiyaci Etüdleri 1955-1956 Yili Neticeleri (Study on drainage requirements in the Yüregir Plain between 1955 and 1956). Tarsus

Devlet Su Isleri (DSI), 6.Bölge Müdürlügü (undated): Protocols No.1-4. Adana

DSI, 6.Bölge Müdürlügü 1982: Asagi Seyhan Sulamasi (Lower Seyhan irrigation). Adana

DSI, 6.Bölge Müdürlügü 1984: A.S.O. IV. Merhale Projesi Yapilirlik Raporu (Planning report for the stage IV project area). Adana

DSI, 6.Bölge Müdürlügü 1986/87: Türkiye'de Büyük Sulama Projelerinin Izlenmesi ve Degerlendirilmesi El Kitabi (Handbook for the evaluation of large irrigation projects in Turkey). Adana

DSI, 6.Bölge Müdürlügü 1988a: 1987 Istatistik Bülteni (Statistical bulletin). Adana

DSI, 6.Bölge Müdürlügü 1988b: Asagi Seyhan Ovasi Sag Sahil Sulamasi Kanallari (Irrigation canals in the right-hand part of the plain). Adana

DSI, 6.Bölge Müdürlügü 1989a: Isletme ve Bakim Calismalari Hakkinda Özet Bilgiler (Information on operation and maintenance activities). Adana

DSI, 6.Bölge Müdürlügü 1989b: Yili Drenaj Kanallari Temizlik Programi (Sag ve Sol Sahil) (Annual drainage cleaning programme). Adana

DSI, 6.Bölge Müdürlügü 1990: Asagi Seyhan Ovasi, IV. Merhale Projesi Planlama Raporu Özeti (Lower Seyhan Plain, Planning report for the stage IV project). Adana

DSI, 6.Bölge Müdürlügü 1991: Drenaj ve Tarla-Ici Gelistirme Projesi. Faaliyet Raporu (1.1.1988 - 31.12.1990) (Drainage and on-farm development project. Activity report I). Adana

DSI, 6.Bölge Müdürlügü 1993: Drenaj ve Tarla-Ici Gelistirme Projesi. Faaliyet Raporu II (1.1.1991 - 31.12.1992) (Drainage and on-farm development project. Activity report II). Adana

DSI, 6.Bölge Müdürlügü 1995a: 1996 Yili Program-Bütce Toplantisi Özet Takdim Raporu (Report on the assessment of the 1996 annual budget). Adana

DSI, 6.Bölge Müdürlügü 1995b: Asagi Seyhan Ovasi Water User Associations. Adana

DSI, 6.Bölge Müdürlügü, ASO Dogankent Isletme ve Bakim Basmühendisligi 1989: 1988 Yili Sulayici Gruplari Calismalari (The Water User Groups' activities in 1988). Adana

DSI, 6.Bölge Müdürlügü, ASO Dogankent ve Yenice Isletme ve Bakim Basmühendisligi 1988: Sulanmayan Parsellerin Sulanmama nedenine göre gruplandirilmasi (Causes for the non-irrigation of plots). Adana

DSI, 6.Bölge Müdürlügü, ASO Ismailiye Isletme ve Bakim Mühendisligi: 1987 and 1988 Yili Sulayici Gruplari Calismalari (The Water User Groups' activities in 1987 and 1988). Adana

DSI, 6.Bölge Müdürlügü, ASO Subesi 1990: Planli Su Dagitimi Uygulama Raporu (Implementation of the water distribution plan), Annex I-IV. Adana

DSI, 6.Bölge Müdürlügü, ASO Subesi 1993: Effects of Drainage and On-Farm Development Works on Groundwater Levels. Adana

DSI, 6.Bölge Müdürlügü, ASO Subesi 1994 (Arca and Donma): Drenaj ve tarla ici gelistirme calismalarinin tabansuyuna etkisi (Impact of drainage and on-farm development works on groundwater conditions). Adana

DSI, 6.Bölge Müdürlügü, ASO Subesi: Asagi Seyhan Ovasi Sulamasi Tabansuyu Kontrol Raporu (Annual control report on groundwater levels and salinity in the Lower Seyhan Plain). Adana

DSI, 6.Bölge Müdürlügü, ASO Yenice Isletme ve Bakim Basmühendisligi: 1987 and 1988 Yili Sulayici Gruplari Calismalari (The Water User Groups' activities in 1987 and 1988). Adana

DSI, 6.Bölge Müdürlügü, Etüd Plan Subesi 1980: Hidroloji Servisi (Hydrological report). Adana

DSI, 6.Bölge Müdürlügü, Isletme ve Bakim Sube Müdürlügü 1989: Sulayici Gruplari ve Faaliyetleri (The Water User Groups and their activities). Adana

DSI, DAPTA, SU-YAPI, TEMELSU, NEDECO (undated): Ana Plan Hazirlanmasi Sulama Stratejisi Irdelenmesi ve Drenaj ve Tarla-Ici Gelistirme Nüve Programi Tasarimi ve Uygulamasi (Research on irrigation strategies, and design and implementation of drainage and on-farm development works). Adana

DSI, DAPTA, SU-YAPI, TEMELSU, NEDECO 1994: Ana (Master) Plan Hazirlanmasi Sulama Stratejisi Irdelenmesi ve Drenaj ve Tarla-Ici Gelistirme Nüve Programi Tasarimi ve Uygulamasi. Proje Tamamlama Raporu, Adana I Alt-Projesi (Research on irrigation strategies, and design and implementation of drainage and on-farm development works. Report on on-going activities, Adana Sub-project I). Adana

DSI, Genel Müdürlügü 1983: The General Directorate for State Hydraulic Works. 1983: Statistical Bulletin with Maps. Ankara

DSI, Genel Müdürlügü 1987: Haritali Istatistik Bülteni 1987 (Statistical bulletin with maps). Ankara

DSI, Genel Müdürlügü 1989: Akim Gözlem Yilligi (Annual discharge measurements). Ankara

DSI, Genel Müdürlügü 1995a: Bakim-Onarim Hizmetleri (Maintenance and repair services). Ankara

DSI, Genel Müdürlügü 1995b: Seyhan irrigation: Net benefits from cotton and wheat. Ankara

DSI, Genel Müdürlügü 1996: The General Directorate for State Hydraulic Works. 1996: Participatory irrigation management in Turkey. (Paper prepared for the „International seminar on participatory irrigation management", 10-17 April 1996). Antalya

DSI, Genel Müdürlügü, 17.Isletme ve Bakim Sube Müdürleri Toplantisi 1993: Sulayici Gurup Faaliyetleri. Annex 1 (The Water User Groups' activities). Izmir

DSI, Genel Müdürlügü, Etüd ve Plan Dairesi Bask. 1988: Asagi Seyhan Projesi. Asagi Seyhan Ovasi Tuzlu ve Sodyumlu Topraklarinin Islahina Yönelik Deneme Raporu (The Lower Seyhan Project. Guidelines for the research project on leaching saline and alkaline soils in the Lower Seyhan Plain). Ankara

DSI, Genel Müdürlügü, Isletme Sube Müdürlügü 1988: Tabansuyu Izleme Calismalari 1979-1986 (Groundwater observations between 1976 and 1986). Ankara

DSI, Genel Müdürlügü, Isletme ve Bakim Dairesi Baskanligi 1995a: Bakim-Onarim Islerinin Kritigi (Criticisms on maintenance and repair works). Ankara

DSI, Genel Müdürlügü, Isletme ve Bakim Dairesi Baskanligi 1995b: Asagi Seyhan Ovasi 10 Yillik Ödenek Dagilimi (Allocation of investments over 10 years in the Lower Seyhan Plain). Ankara

DSI, Genel Müdürlügü, Isletme ve Bakim Dairesi Baskanligi 1995c: Bakim Onarim Ödenekleri (Maintenance and repair). Ankara

DSI, Genel Müdürlügü, Isletme ve Bakim Dairesi Baskanligi (Hasan Özlü) 1993: DSI Sulayici Grubu Calismalari (The DSI's activities with the Water User Groups). Ankara

DSI, Genel Müdürlügü, Isletme ve Bakim Dairesi Baskanligi 1987: Sulama Tesislerinde Tabansuyu Izleme Rehberi (Guidelines for controlling groundwater in irrigation projects). Ankara

DSI, Genel Müdürlügü, Isletme ve Bakim Dairesi Baskanligi 1989: Isletme ve Bakim Bülteni 1988 (Operation and maintenance bulletin 1988). Ankara

DSI, General Directorate, 0&M Department 1995: General information about irrigation works in Turkey and DSI's transfer activities. Ankara

DSI, International Irrigation Management Institute (IIMI) and Economic Development Institute (EDI) 1996: Participatory Irrigation Management in Turkey (Draft), Paper prepared for the International Seminar on Participatory Irrigation Management, Antalya, Turkey, 10-17 April

DSI, Sixth Regional Directorate 1988: Information on DSI Regional Activities. Adana

DSI, Sixth Regional Directorate 1995: Transfer Studies of Irrigation Structure of DSI. Adana

DSI; TOPRAKSU 1961: Lower Seyhan Project Report. Ankara

Dudley, Norman J. 1992: Water Allocation by Markets, Common Property and Capacity Sharing: Companions or Competitors? Natural Resources Journal, Vol.32, pp.757-778

Edelman, Murray J. 1964: The symbolic uses of politics. University of Illinois Press

Elektrik Isleri Etüd Idaresi, Genel Müdürlügü 1989: Türkiye Akarsularinda Su Kalitesi Gözlemleri (Observation stations for the quality control of surface water resources). Ankara

Ergüder, Üstün 1991: Agriculture: The Forgotten Sector, in: Heper, Metin (ed.) 1991: Strong State and Economic Interest Groups. The Post-1980 Turkish Experience. Walter de Gruyter Berlin, New York

Ersoy, Melih 1992: Relations between central and local governments in Turkey: A historical perspective, Public Administration and Development, Vol.12, pp.325-341

Feder, Gershon; Le Moigne, Guy 1994: Umweltverträgliche Wasserwirtschaft (Environmentally sound water supply management). Finanzierung & Entwicklung, Juni

Food and Agricultural Organization (Swayne F. Scott) 1993: Water and Sustainable Agricultural Development. In:DVWK Bulletin 19, pp.19-51. Hamburg, Berlin

Food and Agricultural Organization 1968: Successful Irrigation. Planning, Development, Management. Rome

Food and Agricultural Organization 1975: Water Law in Selected European Countries. Vol.I, pp.211-253 (Turkey). Rome

Food and Agricultural Organization 1987: Preliminary Guidelines for Monitoring and Evaluation of Large-Scale Irrigation Projects in Turkey, TCP/TUR/6652. Rome, Adana

Food and Agricultural Organization 1988: Ex-Post Evaluation of the Lower Seyhan Project (LSP), Vol.II. Rome, Adana

Food and Agricultural Organization 1990: An International Action Programme on Water and Sustainable Agricultural Development. A Strategy for the Implementation of the Mar del Plata Action Plan for the 1990s. Rome

Food and Agricultural Organization 1994: Reforming Water Resources Policy. A Guide To Methods, Processes And Practices. Rome

Food and Agricultural Organization; UNESCO 1973: Irrigation, Drainage and Salinity. An International Source Book. Hutchinson/ FAO/ UNESCO

Frederiksen, Harald D. 1992: Water Resources Institutions. Some Principles and Practices. World Bank Technical Paper No. 191. Washington, D.C.

Frey, Bruno S. 1991: Public Choice. Ergebnisse der letzten zehn Jahre (Public Choice. Results of the last ten years). Wirtschaftswissenschaftliches Studium, Heft 10, pp.492-496

Gardner, Richard L.; Young, Robert A. 1988: Assessing Strategies for Control of Irrigation-Induced Salinity in the Upper Colorado River Basin. American Journal of Agricultural Economics, Vol.70, pp.37-49

Gazi Köy Water User Association 1995: Our Union. Adana

Gibbs, Christopher J.N.; Bromley, David W. 1989: Instutitional Arrangements for Management of Rural Resources: Common-Property Regimes. In: Berkes, Fikret (ed.): Common Property Resources. Ecology and community-based sustainable development. Belhaven Press, London

Goldsmith, D.; Hildyard, N. 1984: The social and environmental effects of large dams. Vol.1/2

Goldsmith; Makin 1989: Canal Lining: From the laboratory to the field and back again. Hydraulics Research ODU OD/P 78

Goodland, Robert; Ledec, George; Webb, Maryla 1989: Meeting Environmental Concerns Caused by Common-Property Mismanagement in Economic Development Projects. In: Berkes, Fikret (ed.): Common Property Resources. Ecology and community-based sustainable development. Belhaven Press, London

Griffin, Ronald C.; Bromley, Daniel W. 1982: Agricultural Runoff as a Nonpoint Externality: A Theoretical Development. American Journal of Agricultural Economics, Vol.64, pp.547-552

Grima, A.P. Lino; Berkes, Fikret 1989: Natural Resources: Access, Rights-To-Use and Management. In: Berkes, Fikret (ed.) 1989: Common Property Resources. Ecology and community-based sustainable development, pp.33-54. Belhaven Press, London

Hardin, Garrett 1968: The Tragedy of the Commons (german version; in G.Hardin, J.Baden (Hrsg.) 1977, Managing the Commons, pp.16-30. San Francisco

Hardin, Russell 1982: Collective Action, in: Resources for the Future. Washington, D.C.

Harsanyi, J.C. 1969: Rational-Choice Models of Political Behaviour vs. Functionalist and Conformist Theories. World Politics 21, pp.513-538

Hartje, Volkmar 1993: Landnutzungsrechte im Sahel (Land use rights in the Sahel region), in: Umweltschutz und Entwicklungspolitik, Schriftenreihe des Vereins für Socialpolitik, Neue Folge Band 226, pp.61-105. Duncker & Humblot, Berlin

Hettich Walter 1975: Bureaucrats and Public Goods. Public Choice, Vol.XXI, pp.15-25

Hilton, J. 1960: Land accumulation in the Turkish Cukurova. Journal of Farm Economics, 42:3

Hilton, Rita M. 1992: Institutional Incentives for Resource Mobilization. An Analysis of Irrigation Systems in Nepal. Journal of Theoretical Politics, Vol.4, No.3, pp.283-308. Sage Publications

Holmstrom, B.R.; Tirole, J. 1989: The Theory of the Firm. In: Richard Schmalensee and Robert Willig, Handbook of Industrial Organization, Vol.1. Elsevier Science Publishers B.V.

Hotes, Frederick L. 1983: The Experience of the World Bank. In: Carruthers, Ian D. (ed) 1983: Aid for the Development of Irrigation. Organisation for Economic Co-Operation and Development. Paris

Hümmer, Philipp 1976: Die Schädlingskatastrophe im Baumwollanbaugebiet der Cukurova/Türkei. Ihre geographischen, wirtschaftlichen und sozialen Konsequenzen (Pest disaster in the cotton farming region of Cukurova/Turkey. Its geographical, economic and social consequences). In: Zeitschrift für ausländische Landwirtschaft, 16.Jg., H.4, pp.372-382

Huppert, Walter; Urban, Klaus 1994: Service Analysis in Irrigation Development. Quarterly Journal of International Agriculture. Vol.33, No.3, pp.260-275

Huppert, Walter; Walker, Hans H. 1988: Management von Bewässerungssystemen. Ein Orientierungsrahmen. (Managing irrigation systems. A Framework) Eschborn

Huppert, Walter; Walker, Hans H.; Wolter, Hans-Werner 1989: Management of Irrigation Systems - A Survey of Concept at GTZ. entwicklung + ländlicher raum. 23. Jg. Heft 2/1989

Hussein, Maliha H. et al. 1987: An Evaluation of Irrigation Projects Undertaken by AKRSP in the Gilgit District of Northern Pakistan. In: International Irrigation Management Institute (IIMI) and Water and Energy Commission Secreteriat (WECS) of the Ministry of Water Resources, Government of Nepal 1989: Public Intervention in Farmer-Managed Irrigation Systems, pp.237-261. Sri Lanka

Hütteroth, Wolf-Dieter 1982: Türkei (Turkey). Wissenschaftliche Länderstudien Band 21. Darmstadt

IBRD 1966: Report of the Seyhan Irrigation Project (Evaluation Team)

IBRD/IDA (Otten and Reutlinger) 1969: Performance Evaluation of Eight Ongoing Irrigation Projects. Economics Department Working Paper No.40. Washington, D.C.

IBRD/IDA 1963: Seyhan Irrigation Project. Adana Province/Turkey

Imar ve Iskan Bakanligi, Planlama ve Imar Genel Müdürlügü, Bölge Planlama Dairesi 1970: Cukurova Bölgesi bölgesel gelisme, sehirlesme ve yerlesme düzeni (Regulation on regional, urban and rural development.) Adana

International Commission on Irrigation and Drainage (ICID) 1967: Controlling Seepage from Canals. ICID Publication. New Delhi/India

International Commission on Irrigation and Drainage 1989: Planning the Management, Operation, and Maintenance of Irrigation and Drainage Systems. A Guide for the Preparation of Strategies and Manuals. World Bank Technical Paper Number 99. Washington, D.C.

International Irrigation Management Institute (IIMI) and Water and Energy Commission Secreteriat (WECS) of the Ministry of Water Resources, Government of Nepal 1989: Public Intervention in Farmer-Managed Irrigation Systems. Sri Lanka

Johnson, Bruce M. 1981: The Environmental Costs of Bureaucratic Governance: Theory and Cases. In: Baden, John; Stroup, Richard L. 1981: Bureaucracy vs. Environment. The Environmental Costs of Bureaucratic Governance. The University of Michigan Press

Jones, W.I. 1992: A Review of World Bank Irrigation Experience. In: LeMoigne, Easter, Ochs, Giltner (eds.), Water Policy and Water Markets: Selected papers and proceedings from the World Bank's Ninth Annual Irrigation and Drainage seminar. World Bank Technical Paper No.249. Washington, D.C.

Kanber, Riza; Yazar, Attila 1992: Effects of saline water table depth on yields of some crops grown in a crop rotation system. Doga 16(1992), pp.287-301

Kanber, Riza; Yüksek, Güner 1979: Cukurova Sulama Rehberi (Guidelines for irrigation). Tarsus Bölge TOPRAKSU Arastirma Enstitüsü, Tarsus

Kasnakoglu, Akder and Gurkan 1990: Agricultural Labor and Technological Change in Turkey, in: Dennis Tully (ed.), Labor and Rainfed Agriculture in West Asia and North Africa, pp.103-133. ICARDA, The Netherlands

Kiratlioglu, A. Esat 1994: Land and water resources management in Turkey. Bari/Italy

Kiray, Mübeccel; Hinderink, Jan 1968: Interdependencies between Agroeconomic Development and Social Change: A Comparative Study Conducted in the Cukurova Region of Southern Turkey. Journal of Development Studies, Vol.4, No.4

Kotschy, Theodor 1858: Reise in den cilicischen Taurus über Tarsus (A journey to the Kilikien Taurus Mountains via Tarsus). Gotha

Köy Hizmetleri 3.Bölge Müdürlügü 1992: Asagi Seyhan Ovasinda Drenaj (Drainage conditions in the Lower Seyhan Plain). Adana

Köy Hizmetleri 3.Bölge Müdürlügü 1993: A.S.O. Developman Calismalari (Proje 1963-1992 Yillari Arasinda). Projenin Yillara Dagilim Uygulama Durum Cizelgesi (Development activities between 1963 and 1992, and their annual distribution). Adana

Köy Hizmetleri 3.Bölge Müdürlügü 1995: 1963-1995 Yillari Arazi Developman Hizmetleri (Development activities between 1963 and 1995). Adana

Köy Hizmetleri Genel Müdürlügü 1990: Sondaj Kuyu Logulari (Groundwater wells). Adana

Köy Hizmetleri Genel Müdürlügü ve Devlet Su Isleri Genel Müdürlügü 1995: Drenaj ve Tarla-Ici Gelistirme Hizmetleri ile ilgili Nüve Programi ve Projelerin Hazirlanmasi, Uygulanmasi. Üc Aylik Is Ilerleme Raporu No.18 (Programme for drainage and on-farm development works, design and implementation. Quadrennial report on ongoing activities No.18). (Dapta, Su-Yapi, Temelsu, Nedeco). Adana, Ankara

Köy Hizmetleri, Genel Müdürlügü 1988: Tarsus Köy Hizmetleri Arastirma Enstitüsü (Research Institute of the General Directorate for Rural Services). Tarsus

Kuzey Yüregir Sulama Birligi 1994 (North Yüregir Water User Association). Adana

Landis, Fred 1994: 'Drainage', Microsoft (R) Encarta. Funk & Wagnalls Coroporation

Leder, Arnold 1979: Party Competition in Rural Turkey: Agent of Change or Defender of Traditional Rule?. Middle Eastern Studies, Vol.15 (1)

Livingston, Marie Leigh 1992: Designing Water Institutions: Market Failures and Institutional Response. World Bank Policy Research Working Paper 1227. Washington, D.C.

Magrath, William 1989: The Challenge of the Commons: The Allocation of Nonexclusive Resources. The World Bank, Policy Planning and Reserach Staff, Environment Department Working Paper No.14. Washington, D.C.

Mann, Charles K. 1972: Formulating a consistent strategy toward on-farm land development in Turkey. United States Agency for International Development, Discussion Paper No.8. Ankara

Martin, Edward D.; Pradhan, Prachanda; Adriano, Marietta S. 1989: Financing Irrigation Services in Nepal. In: Small, Adriano, Martin, Bhatia, Shim, Pradhan 1989: Financing Irrigation Services: A Literature Review and Selected Case Studies from Asia, pp.147-186. Sri Lanka

McKean, Margaret A. 1992: Success on the Commons. A Comparative Examination of Institutions for Common Property Resource Management. Journal of Theoretical Politics 4(3), pp.247-281. Sage Publications

McLean, Ian 1987: Public Choice. An Introduction. Basil Blackwell

Meiri, A.; Shalhevet, J. 1973: Crop Growth under Saline Conditions, in: Arid Zone Irrigation, Ecological Studies 5, pp.277-290. Berlin, Heidelberg, New York

Mertin, W. 1971: Der Wasserpreis in einem Bewässerungsgebiet der Westtürkei (The water prices in a western Turkish irrigation area). Zeitschrift für Bewässerung. Jg.6, H1

Ministry of Agriculture, Forestry and Rural Affairs (MAFRA) 1987: Objectives, Duties and Organizational Structure of the Ministry of Agriculture, Forestry and Rural Affairs. Ankara

Mohamadi, Joma M.; Uskay, Savas 1994: Successful experience with irrigation management through participation and full transfer of management to users in a gradual and an intensive manner. Turkey Case Study. Sofia/Bulgaria

Mueller, Dennis C. 1989: Public Choice II. Cambridge University Press; Cambridge, New York, Port Chester, Melbourne, Sidney

Nabli, Mustapha K.; Nugent, Jeffrey B. 1989: The New Institutional Economics and Its Application to Development. World Development, Vol.17, No.9, pp.1333-1347. Pergamon Press UK

Newman, Katherine S. 1983: Law and economic organization. A comparative study of preindustrial societies. Cambridge University Press; Cambridge, London, New York, New Rochelle, Melbourne, Sydney

Niskanen, William A. 1968: Nonmarket Decision Making. The Peculiar Economics of Bureaucracy. The American Economic Review, Vol.LVIII, No.2, 293-305

Niskanen, William A. 1971: Bureaucracy and Representative Governments. Aldine-Atherton, Chicago

Niskanen, William A. 1973: Bureaucracy: Servant or Master? Lessons from America. Institute of Economic Affairs London

Niskanen, William A. 1978: Competition Among Government Bureaus. In: Buchanan, Rowley, Breton, Wiseman, Frey, Peacock: The Economics of Politics. The Institut of Economic Affairs London

Niskanen, William A. 1981: Preface, in: Baden, John; Stroup, Richard L. 1981: Bureaucracy vs. Environment. The Environmental Costs of Bureaucratic Governance. The University of Michigan Press

Noyan, Kemal et al. 1967: Water Resources in Turkey. International Conference on Water for Peace, Vol.1. Washington, D.C.

Nunns, F. 1956: Work Plans for Soil Survey of Seyhan Delta Area. Turkey. T.A.Project 77-149 USOM/Turkey

O'Mara, Gerald T. 1983: Issues in the Efficient Use of Surface and Groundwater in Irrigation. World Bank Staff Working Papers No.707. Washington, D.C.

Olson, Mancur 1965: The Logic of Collective Action: Public Goods and the Theory of Groups (german edition J.C.B. Mohr Verlag Tübingen 1968)

Onat; Sayin; Yegin 1967: The principles for planning and programming in the development of water resources in Turkey. International Conference on Water for Peace, Vol.6. Washington, D.C.

Organization for Economic Co-Operation and Development (OECD) 1987: Pricing of Water Services. Paris

Ostrom, Elinor 1986a: A Method of Institutional Analysis. In: Kaufmann et al. (eds.), Guidance, Control, and Evaluation in the Public Sector. The Bielefelder Interdisciplinary Project, pp.459-475. Walter de Gruyter Berlin, New York

Ostrom, Elinor 1986b: An agenda for the study of institutions. Public Choice 48, pp.3-35

Ostrom, Elinor 1986c: Multiorganizational Arrangements and Coordination: An Application of Institutional Analysis. In: Kaufmann et al. (eds.), Guidance, Control, and Evaluation in the Public Sector. The Bielefelder Interdisciplinary Project, pp.495-510. Walter de Gruyter Berlin, New York

Ostrom, Elinor 1990: Governing the Commons. The evolution of institutions for collective action. Cambridge University Press

Ostrom, Elinor 1992a: Community and the Endogenous Solution of Commons Problems. Journal of Theoretical Politics, Vol.4, No.3, pp.343-351. Sage Publications

Ostrom, Elinor 1992b: Crafting Institutions for Self-Governing Irrigation Systems. Institute for Contemporary Studies Press, San Francisco, California

Ostrom, Elinor 1992c: Institutions and Common-Pool Resources. Journal of Theoretical Politics, Vol.4, No.3, pp.243-245. Sage Publications

Ostrom, Elinor 1995: Incentives, Rules of the Game, and Development, in: Michael Bruno; Boris Pleskovic (eds.): Annual World Bank Conference on Development Economics 1995, pp.207-234. Washington, D.C.

Ostrom, Elinor; Gardner, Roy; Walker, James 1994: Rules, Games, and Common-Pool Resources. The University of Michigan Press

Ostrom, Elinor; Schroeder, Larry; Wynne, Susan 1993: Institutional Incentives and Sustainable Development. Infrastructure Policies in Perspective. Boulder, San Francisco, Oxford

Özbudun, Ergun 1981: Turkey: The Politics of Clientelism, in: Eisenstadt, S.N.; Lemarchand, Rene 1981: Political Clientelism, Patronage and Development, pp.249-268. Sage Publications Beverly Hills, London

Peacock, Alan 1978: The Economics of Bureaucracy: An Inside View. In: Buchanan; Rowley; Breton; Wiseman; Frey; Peacock: The Economics of Politics

Petermann, Thomas 1993a: Umweltverträglicher Bewässerungslandbau in der Dritten Welt. Welche Rolle kann die UVP spielen? (Environmentally sound irrigated agriculture in Third World countries. What is the role of EIA?). UVP-report 1/1993, 2-7

Petermann, Thomas 1993b, in: GTZ (Deutsche Gesellschaft für Technische Zusammenarbeit) 1993: Irrigation and the Environment. A Review of Environmental Issues. Part I. Eschborn

Pfister, Karl 1971: Rückzahlungsfragen in Bewässerungsprojekten dargestellt an Beispielen aus der Türkei (Repayment problems in irrigation projects illustrated through examples from Turkey). DLG-Verlag Frankfurt am Main

Plusquellec, Herve 1990: The Gezira Irrigation Scheme in Sudan. Objectives, Design and Performance. World Bank Technical Paper No.120. Washington, D.C.

Plusquellec, Herve 1995: Turkey: Transfer of Irrigation Management to Water Users (unpublished)

Pommerehne, Werner W.; Frey, Bruno S. (eds.) 1979: Ökonomische Theorie der Politik (Economic theory of politics). Springer Berlin, Heidelberg

Pommerehne, Werner W.; Frey, Bruno S. 1977: Bureaucratic Behavior in Democracy: A Case Study. Konstanz

Posner, Richard A. 1977: Economic Analysis of Law. Boston, Little, Brown and Company

Postel, Sandra 1992: Last Oasis - Facing Water Scarcity. Worldwatch Institute. The Worldwatch Environmental Alert Studies. New York, London

Pratt, John W.; Zeckhauser, Richard J. (eds.) 1985: Principals and Agents: The Structure of Business. Harvard Business School Press Boston, Massachusetts

Prill-Brett, June 1986: The Bontok: Traditional Wet-Rice and Swidden Cultivators of the Philippines. In: G.G.Marten (ed.), Traditional Agriculture in Southeast Asia. Boulder, Colorado

Quiggin, John 1988: Murray River Salinity - An Illustrative Model. American Journal of Agricultural Economics, Vol.70, pp.635-645

Repetto, Robert 1986: Skimming the Water: Rent-Seeking and the Performance of Public Irrigation Systems. World Resources Institute

Rosen; Windisch; Overdiek 1992: Finanzwirtschaft I. (C) (Public Finance). R.Oldenburg Verlag München Wien

Rowley; Elgin 1985: Towards a Theory of Bureaucratic Behaviour, in: David Greenaway (ed): Public Choice, Public Finance and Public Policy, pp.31-50. Basil Blackwell Oxford /UK

Sagardoy; Bottrall; Uittenbogaard 1982: Organization, operation and maintenance of irrigation schemes. FAO Irrigation and Drainage Paper 40. Rome

Saglamtimur, Timur 1988: Pflanzenproduktionssysteme im Cukurova-Gebiet (Cropping patterns in the Cukurova). C.U., Ziraat Fak. Dergisi. Adana

Saliba, Bonnie Colby; Bush, David B. 1987: Water Markets in Theory and Practice. Market Transfer, Water Values, and Public Policy. Studies in Water Policy and Management, No.12. Westview Press/Boulder and London

Sayin (ulusal calisma grubu) 1993: Türkiye'de Sulu Tarim Yatirimlarina ve Isletme-Bakim Faaliyetlerine Ciftci Katilimi. Inceleme Raporu (International working group: The farmers' contributions towards investments in irrigated agriculture, and towards operation and maintenance activities. Evaluation report). Ankara

Sayin, Sabahattin; Yegin, Hüseyin 1967: The organizations concerned with water resources and problems of coordination in Turkey. International Conference on Water for Peace, Vol.5. Washington, D.C.

Schäfer, Hans-Bernd; Ott, Claus 1986: Lehrbuch der ökonomischen Analyse des Zivilrechts (Textbook for the economic analysis of civil law). Springer Verlag Berlin, Heidelberg, New York, London, Tokyo, Paris

Schaffer, Franz X. 1903: Cilicia. Gotha

Scharpf, Fritz W. 1993: Coordination in Hierarchies and Networks. In: Scharpf (ed.), Games in Hierarchies and Networks. Analytical and Empirical Approaches to the Study of Governance Institutions, pp.125-165. Campus Verlag Frankfurt am Main; Westview Press Boulder, Colorado

Schiffler, Manuel 1993: Nachhaltige Wassernutzung in Jordanien. Determinanten, Handlungsfelder und Beiträge der Entwicklungszusammenarbeit (Sustainable water use in Jordan. Determining factors, action arenas and contributions of development policy). Deutsches Institut für Entwicklungspolitik, Berichte und Gutachten 7/1993. Berlin

Schmitt, Günther 1993: Anforderungen an die Wissenschaft: Was kann die Agrarökonomie von der Institutionen-Ökonomie für die Entwicklung agrar- und umweltpolitischer Konzepte lernen? (Requirements on science: What can agricultural economics learn from the New Institutionalism for the development of agricultural and ecopolitical concepts?) 34. Jahrestagung der Gesellschaft für Wirtschafts- und Sozialwissenschaften des Landbaus. Halle

Schubert, Klaus 1991: Politikfeldanalyse. Eine Einführung (Policy analysis. An introduction). Opladen

Schumann, Jochen 1992: Grundzüge der mikro-ökonomischen Theorie (Principles of micro-economics). Springer Verlag Berlin, Heidelberg, New York

Scutcsh, J.C. 1993: Research and Development for Irrigation and Drainage Maintenance. GRID IPTRID Network Magazine Issue 3

Seger, Norbert 1986: Organisation of the irrigated agricultural economy amongst the small farmers of the Mubaraak/Lower Shabelle region. In: Conze and Labahn (eds.) 1986: Somalia. Agriculture in the Winds of Change, pp.153-163. Saarbrücken-Schafbrücken

Sengoca, C. 1982: The Principal Cotton Pests and their Economic Thresholds in the Kilikien Plain in Southern Turkey. Entomophaga 27 (Special Issue), pp.51-56

Shepsle, Kenneth A. 1989: Studying Institutions: Some Lessons from Rational Choice Approach. Journal of Theoretical Politics, Vol.1, No.2, pp.131-147. Sage Publications

Shresta, Mahesh Man 1987: Problems, Prospects and Opportunities in Developing Farmer-managed Irrigation Systems in Nepal: The Department of Agriculture's Farm Irrigation Program. In: International Irrigation Management Institute (IIMI) and Water and Energy Commission Secreteriat (WECS) of the Ministry of Water Resources, Government of Nepal 1989: Public Intervention in Farmer-Managed Irrigation Systems, pp.201-214. Sri Lanka

Simon, Herbert A. 1957: Models of man, social and rational. Mathematical essays on rational human behaviour in social setting. New York, Wiley

Small, Leslie 1987: Irrigation Service Fees in Asia. ODI/IIMI Irrigation Management Network Paper 87/1c. London: Overseas Development Institute

Small, Leslie E. 1989: Financing Irrigation: A Literature Review. In: Small, Leslie E., et al. 1989: Financing Irrigation Services: A Literature Review and Selected Case Studies from Asia, pp.3-23. Sri Lanka

Small, Leslie E.; Adriano, Marietta S. 1989: Financing Irrigation Services in Indonesia. In: Small, Leslie E., et al. 1989: Financing Irrigation Services: A Literature Review and Selected Case Studies from Asia, pp.27-64. Sri Lanka

Small, Leslie E.; Adriano, Marietta S. 1989: Financing Irrigation Services in the Philippines. In: Small, Leslie E., et al. 1989: Financing Irrigation Services: A Literature Review and Selected Case Studies from Asia, pp.189-228. Sri Lanka

Small, Leslie E.; Adriano, Marietta S.; Martin, Edward D.; Bhatia, Ramesh; Shim, Young Kun; Pradhan, Prachanda 1989: Financing Irrigation Services: A Literature Review and Selected Case Studies from Asia. Sri Lanka

Small, Leslie E.; Carruthers, Ian 1991: Farmer-Financed Irrigation. The Economics of Reform. Wye Studies In Agricultural And Rural Development. Cambridge University Press

Small, Leslie E.; Svendson, Mark 1992: A Framework for Assessing Irrigation Performance. International Food Policy Research Institute, Working Papers on Irrigation Performance 1. Washington, D.C.

Smout, I. 1990: Farmer Participation in Planning, Implementation and Operation of Small Scale Irrigation projects. ODI-IIMI Irrigation Management Network Paper 90

Soysal, Mustafa 1976: Die Siedlungs- und Landschaftsentwicklung der Cukurova mit besonderer Berücksichtigung der Yüregir-Ebene (Regional and settlement development of the Cukurova with special considerations given to the Yüregir Plain). Erlanger Geographische Arbeiten (Vorstand der Fränkischen Geographischen Gesellschaft), Sonderband 4. Erlangen

Soysal, Mustafa 1985: Zur Organisation der Wanderarbeiter im türkischen Baumwollgebiet der Cukurova (Organization of migrant labourers in the Turkish cotton area Cukurova). Berichte über Landwirtschaft, Band 63, pp.162-168

Soysal, Mustafa; Yurdakul, Oguz 1986: Adana Ovasi Pamuk Isciligi (Cotton labourers in the Adana Plain). Adana

Stroup, Richard L.; Baden, John A.; Fractor, David T. 1983: Natural Resources. Bureaucratic Myths and Environmental Management. Pacific Institute for Public Policy Research. San Francisco, California. Cambridge, Massachusetts

Subramanian, Ashok; Jagannathan, N.Vijay; Meinzen-Dick, Ruth 1995: User Organizations For Sustainable Water Services. World Bank Water Resources Seminar (Chantilly, Virginia, December 11-13, 1995)

Sulama Tesisleri Isletme ve Bakim Hizmetlerinin Köy Tüzel Kisiliklerine Devri Hakkinda Sözlesme (Contract for the Water User Groups' contributions to O&M services). Adana

T.C. Bayindirlik ve Iskan Bakanligi; Devlet Su Isleri Genel Müdürlügü: 1987, 1988, 1989, 1990 Yili Sulama ve Kurutma Tesisleri, Isletme-Bakim ve Yillik Yatirim Ücret Tarifeleri (Annual Decree on Water Tariffs for O&M and Investment). Ankara

T.C. Ministry of Energy and Natural Resources, General Directorate for State Hydraulic Works 1980: Lower Seyhan Basin Master Plan. Vol.III. Ankara

T.C. Prime Ministry, State Planning Organization 1987: Fifth Five Year Development Plan 1985-1989. Ankara

T.C. Prime Ministry, State Planning Organization 1989: The Southeastern Anatolia Project Master Plan Study, Vol.2. Tokyo, Ankara

T.C. Resmi Gazete (Law Gazette), 1924, Village Law No.442

T.C. Resmi Gazete (Law Gazette), 1953, Law No.6200

T.C. Resmi Gazete (Law Gazette), 1960, Law No.7457

T.C. Resmi Gazete (Law Gazette), 1981, Law No.293.4/101

T.C. Resmi Gazete (Law Gazette), 1985, Law No.3202

T.C. Resmi Gazete (Law Gazette), 1988, Law No.19919

T.C. Resmi Gazete (Law Gazette), 1989, Law No.20106

T.C. Resmi Gazete (Law Gazette), Civil Law No.743

Taylor, M 1987: The Possibility of Cooperation. Cambridge University Press. Cambridge

Tekinel, Osman 1992: Cevre-Etki Degerlendirilmesi (CED) Yönünden Asagi Seyhan Ovasi Sulamasi Sorunlari ve Cözüm Önerileri (Environmental Impact Assessment for irrigation in the Lower Seyhan Plain, and proposals for solution). Adana

Tekinel, Osman; Dinc, Gürol 1975/76: Development of Irrigation and Drainage Works in the Lower Seyhan Plain. Adana

Tekinel, Osman; Kanber, Riza 1988: Results of cotton irrigation experiments carried out by the University of Cukurova. Adana

Tekinel, Osman; Kanber, Riza 1990: Importance of using experiences gained from the Lower Seyhan Irrigation Project. Bari/Italy

Tekinel, Osman; Yazar, Attila 1984: General Discription of the Agricultural Situation in the Province of Adana. Environmental Health Impact Assessment Workshop. Adana

Tekinel; Dinc, G.; Kumova 1976a: Groundwater status of the Lower Seyhan Plain Irrigation Project Area and its anticipated problems. Adana

Tekinel; Dinc, G.; Kumova 1976b: Asagi Seyhan Sulama Proje Alaninda Tarla Ici Drenajina Neden Olan Yüksek Taban Sorunu Üzerinde Bir Inceleme (On-farm drainage systems, and causes for high groundwater levels in the Lower Seyhan Irrigation Project). Adana

Tekinel; Kanber; Özbek; Drahor; Kanalici; Yalbuzdag; Tetik; Toyganözü; Gencay; Cataloglu 1993: Türkiye'de Sulu Tarim Yatirimlarina Ciftci Katilimi ve Geri Ödeme (Cukurova Örnegi) (The farmers' contributions to investments in irrigated agriculture, and repayment problems in Turkey: the Cukurova). Adana

Tekinel; Kumova; Dinc; Yazar; Kanalici; Kirmaci 1985: Asagi Seyhan Sulama Proje Alaninda Son Yirmi Yilda Taban Suyu Düzeyi Degisimleri (Report on groundwater levels in the Lower Seyhan Irrigation Project for the last twenty years), in: II. Ulusal Kültürteknik Kongresi. Adana

The Majelle, in the english translation of Majallahel-Ahkam-I-Adliya. Translated by C.R.Tyser (President of the District Court of Kyrenia), D.G.Demetriades (Registrar of the District Court) and I.H. Effendi (Turkish Clerk at the same court). Lahore/Pakistan

The World Bank 1978: Turkey. Seyhan Irrigation Project Stage II. Completion Report

The World Bank 1983: World Development Report 1983. Oxford University Press. New York

The World Bank 1985: Impact Evaluation Report Turkey - Seyhan Irrigation Project (Stage II), Loan 587-TU, Credit 143-TU), March 12

The World Bank 1992: Turkey. Irrigation Management and Investment Review

The World Bank 1993: Water Resources Management. A World Bank Policy Paper. Washington, D.C.

The World Bank 1994a: A Review of World Bank Experience in Irrigation. Report No.13676. Operations Evaluation Department. Washington, D.C.

The World Bank 1994b: World Develoment Report 1994 (Infrastructure for Development). Oxford University Press, New York

The World Bank; ICID 1989a: Research and Development in Irrigation and Drainage. (Draft)

The World Bank; ICID Dec. 1989b: Irrigation and Drainage Research. Vol.I, Working Draft

Tietzel, Manfred 1981: Die Rationalitätsannahme in den Wirtschaftswissenschaften oder Der homo oeconomicus und seine Verwandten (The rationality approach in economics, or the homo oeconomicus and his relatives), in: Jahrbuch für Sozialwissenschaft, Bd. 32

Tirole, Jean 1988: The Theory of Industrial Organization, pp.51-55. The MIT Press Cambridge, Massachusetts; London, England

Tollison, Robert D. 1982: Rent-Seeking: A Survey. Kyklos, Vol.35, pp.575-602

TOPRAKSU Genel Müdürlügü 1974: Seyhan Havzasi Topraklari (Havza No.18) (Soils in the River Seyhan basin), Raporlar Serisi 70. Ankara

Tsur, Yavoc; Dinar, Ariel 1995: Efficiency and Equity. Considerations in Pricing and Allocating Irrigation Water. Policy Research Working Paper 1460, The World Bank. Washington, D.C.

TU Berlin 1992: Pestizideinsatz in Bewässerungsgebieten der Südtürkei/Cukurova (Pesticide use in irrigated areas in South Turkey/Cukurova). Berlin

Tug, Aydin 1975: Village Administration in Turkey. T.C. State Planning Organization Pub. No. 1421 - CD: 291. Ankara

Tullock, Gordon 1965: The Politics of Bureaucracy. Public Affairs Press, Washington, D.C.

Tunc, Ahmet et al. 1979: Cukurova'da Pamukta Bulunan Zararli Ve Faydali Böceklerin Populasyonlari Üzerinde Arastirmalar (Research on cotton pests in the Cukurova), Bölge Zirai Mücadele Arastirma Enstitüsü, pp.181-195. Adana

Turfan; Bozkus 1992: Hydropower potential in Turkey. Water Power&Dam Construction, December 1992, pp.11-12

Turral, Hugh 1995: Recent trends in irrigation management - changing directions for the public sector. ODI Natural Resource *perspectives*, No.5. London

U.S. A.I.D. Project Evaluation Report No.50 1983: On-Farm Water Management In Agean Turkey, 1968-1974

Ulusan, Aydin 1980: Public Policy Towards Agriculture and Its Redestributive Implications, in: Özbudun and Ulusan (eds.): The Political Economy of Income Distribution in Turkey. Holmes & Meier Publishers, London

Uner, Naki 1967: Problems of Irrigation Development in Turkey and Improvement of the Ways of Solution. International Conference on Water for Peace, Vol.7. Washington, D.C.

States Department of Agriculture (USDA) 1954: Diagnosis and Improvement of Saline and Alkali Soils. Agriculture Handbook No.60. Washington, D.C.

United States Man and Biosphere Program/United States Agency for International Development (Robert E. Tillman) 1981: Environmental Guidelines for Irrigation. New York Botanical Garden Cary Arboreum

Ünver, Olcay et al. 1994: Identification of management, operation and maintenance model for irrigation systems in the Southeastern Anatolia Project (GAP) of Turkey. Ankara

Uphoff, Norman; Ramamurthy, Priti; Steiner, Roy 1991: Managing Irrigation: Analyzing and improving the performance of bureaucracies. New Delhi/India

Van Tuijl, Willem (undated): Improving Water Use in Agriculture. Experiences in the Middle East and North Africa. World Bank Technical Paper No.201. Washington, D.C.

Wade, Robert 1982: Irrigation and Agricultural Politics in South Korea. Westview Press Boulder/Colorado

Wade, Robert 1987: Managing Water Masters: Deterring Expropriation or Equity as a Control Mechanism. In: Water and Water Policy in World Food Supply, pp.117-183. A&M University Press, Texas

Wade, Robert 1988: Village Republics. Economic conditions for collective action in South India. Cambridge University Press, Cambridge, New York, New Rochelle, Melbourne, Sidney

Wade, Robert 1993: The Operation and Maintenance of Infrastructure: Organizational Issues in Canal Irrigation. Sussex/England

Walker, Hans; Cleveringa, Rudolph 1993: Management of Irrigation Systems. Working Aids for Irrigation System Management. Vol.I, II. Deutsche Gesellschaft für Technische Zusammenarbeit (GTZ). Eschborn

Weber, Max 1947: The Theory of Social and Economic Organization. New York: The Free Press of Glencoe

Weigel, Wolfgang 1987: Zur ökonomischen Analyse öffentlicher Institutionen (Economic analysis of public institutions). Veröffentlichungen der Kommission für Sozial- und Wirtschaftswissenschaften, Wilhelm Weber (ed.), Nr.24. Verlag der österreichischen Akademie der Wissenschaften. Wien

Weiker, Walter F. 1981: The Modernization of Turkey. Holmes&Meier Publishers New York, London

Weissing, Franz; Ostrom, Elinor 1993: Irrigation Institutions and the Game Irrigators Play: Rule Enforcement on Government- and Farmer-Managed Systems. In: Fritz Scharpf (ed), Games in Hierarchies and Networks. Analytical and Empirical Approaches to the Study of Governance Institutions, pp.387. Frankfurt am Main, Boulder/Colorado

Werner, Frank 1995: Internationale Erfahrungen mit Wassermärkten. Eine Analyse unter Hervorhebung der Wassermärkte Arizonas und Chiles (International experiences with water markets, with particular reference to the water markets in Arizona and Chile). Berlin

Williamson, Oliver E. 1979: Transaction Cost Economics: The Governance of Contractual Relations. Journal of Law and Economics 22(2), pp.233-261

Wohlwend, Bernard J. 1981: From Water to Water Resources Law and Beyond. Water International, 6(1981), pp.2-15

Wolff, Peter 1987: Stabilisierungspolitik und Strukturanpassung in der Türkei 1980-1985 (Stabilization policies and structural adjustment in Turkey between 1980 and 1985). Berlin

Wolff, Peter; Le Claire, Bertil; Krüger, Axel; Naumann, Heiner; Scheld, Friedemann 1983: Die Entwicklung der Nahrungsmittelindustrie in der Cukurova-Region/Türkei (The development of the food-stuffs industry in the Cukurova region). Berlin

Yan Tang, Shui 1991: Institutional Arrangements and the Management of Common-Pool Resources. Public Administration Review, Vol.51, No.1, pp.42-50

Yan Tang, Shui 1994: Institutions and Performance in Irrigation Systems, in: Ostrom et al. 1994: Rules, Games, and Common-Pool Resources, pp.225-246. The University of Michigan Press

Yan Tang, Shui; Ostrom, Elinor 1993: The Governance and Management of Irrigation Systems: An Institutional Perspective. ODI Irrigation Management Network Paper 23

Yarpuzlu; Dogan 1986: Tarsus Ovasi Kapali Drenaj Projeleme Kriterleri (Criteria for closed drainage systems in the Tarsus Plain). Köy Hizmetleri Genel Müdürlügü/Tarsus Arastirma Enstitüsü Md., No.65. Tarsus

Yeniceri, Cengiz 1980: Asagi Seyhan Projesi Ciftci Egitim Servisi Calismalari (The Agricultural Extension Service in the Lower Seyhan Project). Adana

Yurdakul, Oguz 1992: Güneydogu Anadolu Projesinin Tamamlanmasinin Cukurova Bölgesi Tarimina Olasi Etkileri (The impact of the Southeastern Anatolian Project on the agricultural sector in the Cukurova), in: Gap'in Cukurova Tarimina Muhtemel Etkileri, Paneli Konusmalari, pp.5-11. Adana, Ankara

Yurdakul, Oguz; Ören, Necat M. 1991: Cukurova Bölgesinde Pamuk Üretim Maliyeti Satis Fiyati ve Ekim Alani Iliskisi (Relation between cotton production costs, and the area planted with cotton), in: Cukurova 1. tarim kongresi, pp.32-41. Adana

Index

Printing: Weihert-Druck GmbH, Darmstadt
Binding: Buchbinderei Schäffer, Grünstadt